CREATING AND MAINTAINING

A WORLD-CLASS MACHINE SHOP

A Guide to General and Titanium Machine Shop Practices

BY

EDWARD F. ROSSMAN

INDUSTRIAL PRESS INC.
NEW YORK

Library of Congress Cataloging-in-Publication Data
Rossman, Edward F. Creating and maintaining a world-class
machine shop / Edward F. Rossman
p. cm.
ISBN 0-8311-3299-X
ISBN 978- 0-8311-3299-6

1.machine shop , etc. I. Title.

TS183.B73 2007   670–dc22

Creating and Maintaining a World-Class Machine Shop:
A Guide to General and Titanium Machine Shop Practices

First Edition
2007

Managing Editor: John Carleo
Copy Editing: Robert Weinstein
Interior Text and Cover Design: Janet Romano

Copyright © 2007 by Industrial Press Inc., New York.
Printed in the United States of America. All right reserved.
This book, or any parts thereof, may not be reproduced, stored in a retrieval system,
or transmitted in any form with-out the permission of the publisher.

10   9   8   7   6   5   4   3   2   1

TABLE OF CONTENTS

Acknowledgements — vii
Preface — ix
On This Book's Arrangement — xiii
Introduction — xv

Group I: Defining a World-class Machine Shop

Chapter 1: Traits of a World-Class Machine Shop — 1
Key Characteristics or Traits of that Ideal World-Class Shop: — 1
Examples and Discussion of World-Class Elements in Shops — 6

Chapter 2: A 30-Minute Shop Assessment — 9
The Shop Tour — 10
Check List: Shop Assessment — 17

Chapter 3: Selection and Training of Employees — 19
Leadership and Employee Involvement — 20
NC Programming — 21
Lean Studies — 22
Summary — 23

Group II: Equipment — 25

Chapter 4: Work Holding and Fixturing — 27
Fixture Concepts that Improve Efficiency — 27
Setup Time Considerations — 35
Fixture Materials — 36

Chapter 5: Machines — 39
Long Range Plans/Strategy/Mission — 39
Machine Location and Future Re-location — 41
Part Requirements — 42
Defining the Machine — 43
Maintenance and Preventive Maintenance — 57
Process Considerations — 58
Future Machines — 58

Group III: Milling of Titanium alloys — 61
Definitions — 62

Chapter 6: Rough Milling of Titanium alloys — 63
Traditions and Basics for Rough Milling of Titanium Alloys — 63
Chatter—A Primary Limiter to Metal Removal When Roughing — 67
Efficient Rough Milling Today — 67
Summary — 71

Chapter 7: Finish Milling of Titanium alloys — 73
Traditions and Basics for Finish Milling of Titanium Alloys — 73
Efficient Finish Milling Today — 74
Extending Cutter Life — 77
Summary — 77

Chapter 8: Maintaining Large Part Accuracy — 79
Temperature Control and Coolants — 80
Statement of the Problem — 80
Drill Jig Accuracy and Limitations — 81
Development of Precision Drilling Centers (PDCs) — 85
Finding Other Precision Machines to Handle Production Overload — 85
Development of a Test Fixture to Predict Machine Capability — 85
Results and Findings of Testing — 87

Chapter 9: Milling Difficult Features — 89
Fundamental Milling Parameters for Titanium alloys — 89
Milling Thin Webs in Titanium alloys — 91

Milling Thin Flanges in Titanium Alloys	93	**Chapter 15: Cutter Life**	**129**
How to Produce a Rough Surface Finish	95	General Notes on Cutters	130
Summary	96	Parameters for Finish Milling With Solid Carbide Cutters	132
Chapter 10: Warpage	**97**	Finish Milling Costs	133
Problem	97	Summary	133
Purpose	97		
Coverage and Limitations	98	**Group VI: Efficiency and Lean Practices**	**135**
		Chapter 16: Inter-Company Sharing	**137**
Group IV: Sanding and Vibratory Finishing	**105**	Problem	138
Chapter 11: Sanding and Benching	**107**	Why Lean is Not Enough	139
Problem	108	Examples of Success From Sharing	139
Argument for Changing the Specification	109	Allay Fears of Giving Away Ideas and Sharing	141
Arguments Against Changing the Specification	109	Summary	141
Details and History on Growth in Costs of Milling	110	**Chapter 17: Continuous Improvement Is Not Enough**	**143**
Conclusions	110	Potential Areas for Rapid Improvement	144
Chapter 12: Vibratory Finishing	**113**	Examples of Rapid Advances in Milling	145
Vibratory Finishing is Used Overseas	114	Summary	145
Early Discussions With Machine Houses	114	**Chapter 18: Lean Studies & Communication**	**184**
Vibratory deburring equipment:	114	Some Examples of Quantum Savings	184
Part Fixturing	114	Process Stabilization	186
Media for Finishing	115	Many Companies Do Not Use Lean	186
OSHA Considerations for Media and Liquids	115	Summary	151
Support Team	115	**Chapter 19: Measuring Shop Performance**	**153**
Summary	116	My Recommendations for Performance Measurement	154
Group V: Cutters	**117**	What To Do With the Data	157
Chapter 13: Sources of Cutters	**119**	Speed Up the Metal Cutting Time	157
Standard Cutters	119	The Value-Added Operations:	157
Specialized Cutters	121	Summary	158
Tool Holders and Balancing	122	**Chapter 20: Work Cells**	**159**
Summary	122	Transformation to Work Cells	160
Chapter 14: Cutter Checking	**123**	Demise of Early Work Cells	162
List of Cutter Checking Practices	124	The Beginnings of Assembly Lines	162
Ideas for Monitoring Cutter Wear	126	Work Cells Re-emerge	162
Summary	127		

Flexible Work Cells	163
Today We Have Lean Studies	164
Group VII: Ed's Tool Kit	**165**
Chapter 21: Tidbits and Rules of Thumb	**167**
Safety Issues	167
Rules of Thumb	168
Potential Resources for Information Research on Titanium Alloy Milling	169
Chapter 22: My Tales Memorable Happenings	**171**
Hot Equipment Quits Working	171
Minds Have Great Power	171
Sometimes We Jump Too Soon	172
Sometimes We Outsmart Ourselves	172
Index	*173*

LIST OF ILLUSTRATIONS

Figure 4-1:	Rotisserie Fixtures	29
Figure 4-2:	Single Rotary Fixture	30
Figure 4-3:	Dovetail Fixture	31
Figure 4-4:	Pogo Tool	32
Figure 4-5:	Sketch of Tooling Tab Use	32
Figure 4-6:	Sketch of Part with Clamping Grooves	33
Figure 4-7:	Sketch of Tabs, Bonded or Welded	34
Figure 6-1:	Cutter Life Curve For Rough Milling	65
Figure 6-2:	Cutter Life Curve For Rough Milling (Metric Values)	66
Figure 6-3:	Plunge Roughing	68
Figure 6-4:	Trochoidal Milling	69
Figure 9-1:	Minimum Flange Thickness vs. Height for Titanium 6AL-4	94
Figure 10-1:	Straightening by Shot Peening	101
Figure 11-1:	Sanded Surface, Scratches Down	109
Figure 11-2:	Milled Surface, Points Up, Between Passes	110
Figure 13-1:	Telescoping Heat Shrink Sections	121
Figure 14-1:	Fail-Safe Pin on Fixture	125
Figure 15-1:	Minimized Rough Milling Costs	130
Figure 15-2:	Finished Milling Cutter Life	132
Figure 15-3:	Finished Milling Cutter Life (Metric Values)	133
Figure 19-1:	Quantity of Observations for Given Confidence Levels	156
Figure 19-2:	Delay Study Results Example	156
Figure 20-1:	Robbins & Lawrence Armory & Factory, Windsor, Vermont (photoby the author)	161

DEDICATIONS

When I was about ten years old, a car passed us with the rear window covered with stickers of U.S. National Parks that had apparently been visited by the unknown driver. I mentioned to my father how exciting it would be to have visited all those places. My father answered, **"It is more important to be interested in where you are going than in where you have been."** My father, Seward A. Rossman, is my key inspiration. While I was in college and early in my career he challenged me to become the best engineer possible and furnished many ideas on making a shop more flexible and functional at very low cost. I have included his ideas in this book, and I dedicate this book to him.

ACKNOWLEDGEMENTS

Further Credits and Acknowledgements

Please note that this book is a collection of ideas toward creating world-class machine shops and toward more efficient milling of titanium alloys. I can not remember exactly where each bit of information came from, but the following persons come to mind: Paul Schaffner of GKN St. Louis, Jonathan Saada of Hanita Cutting Tools, and Hector Davis of Forrest Machine for fresh efforts with *powdered metal cutters*. Keith A. Young of Boeing Phantom Works for thoughts on *milling chatter* and on *spindle growth* (2004), Mike Watts and collaborators of Boeing Commercial MR&D who in 1997 launched an enterprise-wide "*High performance titanium research*" effort, Jonathan Saada of Hanita Cutting Tools for his continuing efforts in reducing the *costs of milling titanium alloys*, David West of MR&D of Boeing for his continuing and successful efforts on *high speed milling of titanium alloys*, John Cave of Aerospace Dynamics International for his thoughts on *progressive management style*, and Hector Davis (formerly of Aerospace Dynamics International) who implemented and improved on our efforts to increase the *speed of finish milling* of titanium alloys, also Garry Booker of Boeing, Kevin Van Dyke and Tom Hoffman of Summit Design & Manufacturing for great pioneering work in rough milling (particularly *plunge roughing*) of titanium alloys. I also give my thanks and appreciation to those not remembered and not acknowledged by this aging mind.

My thanks to Harold Bloom, author of many books including *Genius*, from which I borrowed the idea of grouping the chapters within this book. Thanks also to my granddaughter Katie McFadden (working on her Master's in English at Portland State University) for her suggestions and editing help, my daughter Teresa McFadden for her typing and editing help, and to John C. Dickson, my manager, for his critiques and insights into my writing.

My greatest thanks are to my wife, Iolanda, who has kept me alive for the last 47 years with her love, care, and patience.

PREFACE

This book is an attempt to cover many of the manufacturing aspects involved in creating and maintaining a world-class machine shop. While the primary focus of this book is on milling titanium alloys, many of the items discussed could apply to any machining or manufacturing enterprise. The look and feel of a world-class shop is described with many examples. Selection, education, and training of manufacturing people are discussed. Features of shop equipment including machines and fixturing are covered. The latest information available on efficient milling of titanium alloys is described, and milling of difficult features is covered along with recommended courses of action to minimize warpage and part distortion. Benching and sanding of parts is discussed because these are expensive steps in the fabrication of titanium alloy parts. Cutters and cutter coatings are covered. Efficient and lean practices are described briefly along with some frugal metrics for measuring progress. My thoughts on inter-company sharing of machining technology are championed. Tidbits of useful information from my toolbox and memorable experiences from my world of milling are also included.

Why This Book?

Many machining suppliers have asked for the information in this book, and my primary motivation is a quest to minimize the costs of milling titanium alloy. The purpose of this book is:

To provide information on milling of titanium alloys in anticipation of milling suppliers needs.

To document information that will result in the reduction of the cost of milling titanium alloy parts so:

Customers can afford more products.

We can extend the time that machining stays in this country, hopefully forever.

The notion that we machinists can become more competitive in this world market is not a bad reward for my efforts.

My Authority

I have had a life-long interest in machining. After growing up near Detroit and attending General Motor's Institute (now Kettering University), I began my machining career at General Motor's Truck in Pontiac for nine years. In 1962 I moved to Seattle and worked with Boeing on several space projects including the Saturn Moon Rocket. I then switched to shipbuilding for four years with Lockheed in Seattle. I helped fabricate computer peripheral equipment with Tally Corp. for twelve years, and then shifted to Sundstrand to build aircraft black boxes for four years. In 1983 I moved to Detroit for five years to help Focus:HOPE (a human and civil rights organization) launch and administer a machinist-training program for minorities. This was an opportunity to give back to the community that has been so good to me. I returned to Seattle in 1989 and was promoted to Associate Technical Fellow in Manufacturing R&D with Boeing Integrated Defense Systems. I am considered an expert in machining and thermal processing. I received my Ph.D. in engineering management at age 66, and am currently a consultant and troubleshooter with machining suppliers for Boeing. I have published and presented several research papers on machining and thermal processing of titanium alloys through the Society of Manufacturing Engineers.

I am a past chairman of the Machining Technology Association of the Society of Manufacturing Engineers, and am a current steering committee member of the Machining and Material Removal community of SME.

On 22 January 2004 at the Complex Machining Symposium in Valencia, California, I was awarded an "Industry Outstanding Contribution Award" by Hanita Cutting Tools, "For his long-time promotion and education of high speed titanium machining to the machining industry."

I also received the "2005 Outstanding Member Award" from the Society of Manufacturing Engineers (SME) on August 3 during the SME technical summit. "For Ed's tireless work on helping countless suppliers become more proficient on titanium machining and for helping the SME organization grow with his involvement in the Machining and Material Removal Community." The above quote is from my supervisor John C. Dickson.

I and Iolanda, my wife of 47 years, live in Auburn, Washington and our six grown children and eight of our nine grandchildren live within an hour's drive. Iolanda and I do a lot of traveling together as part of my work and to visit our other granddaughter working on her master's degree in English at Portland State University, and we visit family and friends in Fresno, New York, Detroit, Montana, California, and St. Louis.

Why this Material Should Interest You

This book is about proven principles of success in metal fabrication. The material presented is practical and in use, as the examples will show. Theory is not discussed except in those areas where I attempt to predict what may transpire next.

I am addressing "you":

>Titanium machining shops
>General machine shops
>And to some degree, all fabrication shops

What to Do With this Information?

Machining suppliers have indicated that they want to be the best—to be world-class and to excel in this highly-competitive world of machining. I also expect that if I were to visit your shop, I would find that you are doing most of the good actions that are discussed in this book. But, I have attempted to provide some additional steps that you can take to become world-class and to maintain that status.

If you have some further thoughts on becoming or on what it means to be a world-class shop that you are willing to share with me and with others, I want to hear them. I can be contacted by email at.

http://www.industrialpress.com/en/Home/tabid/36/bookdetails/tabid/54/CatalogItemID/395/Default.aspx. I do not consider this to be the last word on this matter. In fact, to quote a former manager, Linde Criddle, "The only reason we write things down is so we have a basis for changing them." He said this in response to my complaint that I had just finished a technical report only to receive further pertinent and important inputs.

Take out your pen and write notes in this book for your own purposes because many elements in this book will be obsolete the moment it is published. Two of the reasons for this anticipated obsolescence are:

We keep getting smarter at milling and at finding out about progress that others have made, and

Technology is advancing at a screaming pace.

And this, too, shall pass away:

> It is said an Eastern monarch once charged his wise men to invent him a sentence to be ever in view, and which should be true and appropriate in all times and situations. They presented him the words: **'And this, too, shall pass away.'**
>
> — Abraham Lincoln, September 30, 1859

ON THIS BOOK'S ARRANGEMENT

The chapters in this book are placed within seven groups.

Group I: Defining a World-Class Machine Shop attempts to define a world-class shop for milling titanium alloys—then takes you for a simulated shop tour on how to assess your shop in world-class measures. Most important, there is a discourse on selection and training of people, including management. Furthermore, I have included the checklist I use for a 30-minute shop assessment.

Group II: Equipment discusses fixtures (work holding devices) and the metal cutting machinery. How does one equip such a world-class titanium machine shop? The challenge with fixtures is to minimize costs and the number of fixtures and make them easy to use. Machinery selection, maintenance, and use are also discussed.

Group III: Milling of Titanium Alloys covers the bulk of value-added milling operations—metal removal. The sequence within this group begins with rough milling, and then proceeds to finish cuts. These chapters are followed with thoughts on maintaining the accuracy of large parts. Some trials and tribulations of milling difficult part sections are covered. The chapter on warpage discusses the three causes of warpage with examples of how to prevent such distortion.

Group IV: Hand Finishing and Vibratory Finishing discusses benching in two chapters. The last step in fabrication is benching—final sanding and deburring to achieve the specified surface finish which follows the normal sequence of machine shop milling.

Group V: Cutters is intended to help you decide on sources of cutters and cutter holders, parameters for checking the accuracy of cutter/tool holder setups, assuring accuracy of cuts, detecting cutter wear, and determining cutter life. Cutters are an important element in milling and warrant a separate grouping.

Group VI: Efficiency and Lean Practices is about keeping costs low and getting product out quickly. Chapters in this group discuss a rather new topic—Intercompany sharing of technology. This is the notion that continuous improvement is not enough; other factors are also important such as lean studies and communication, metrics for determining shop health, a discussion of work cells vs. other groupings of equipment, and the importance and ways to be flexible—being able to rapidly relocate and rearrange equipment.

Group VII: Ed's Tool Kit is a collection of items that I use in trying to make all of the above actions happen. I initially labeled this group as miscellaneous, but it is much more than just a collection of items. This group of notes includes: tidbits, safety issues with titanium, memorable happenings, and other items from my notebooks that I find helpful on a daily basis and believe are worth sharing with you.

INTRODUCTION

What makes development of a world-class shop possible and necessary? The spirit of the age—the times we live in.

This book contains examples of world-class elements from many shops that I have visited. Being world-class is certainly not a cut and dried thing, but the elements included are based on my judgment of what is important for competition in this world of ours, and is an attempt to provide criterion for evaluation of machine shops.

Historical happenings in the world of machining and manufacturing have played a major role in my collection of world-class traits for machine shops. I have visited several places of greatness in fabrication including Henry Ford's Highland Park, Michigan, plant. No, it wasn't running when I visited, but it still smells like a factory. I have visited the Precision Museum [of Machining] in Windsor Vermont—the birthplace and museum of many of the world's precision machinery concepts. I visit and read about places of greatness in our field of machining and manufacturing to see the roots of our machining and manufacturing technology.

In this book, I venture upon a purely personal definition of world-class as it relates to shops that machine airframe parts with a focus on milling titanium alloys. I base this book, *Creating and Maintaining a World-Class Machine Shop,* upon my belief that understanding and appreciating elements of world-class fabrication is the best mode for achieving and maintaining a great shop.

I dream of inventing great machine shops. Emerson, in his essay "Quotation and Originality," slyly observed, "Only an inventor knows how to borrow." This book is packed with borrowed ideas that I have attempted to package and organize in a manner that I hope will augment your dreams. Harold Bloom, in his book *Genius* states, "To be augmented by the genius of others is to enhance the possibilities of survival, in the present and near future." I believe that the opportunities for shop improvement (your survival) know no limits.

GROUP I: DEFINING A WORLD-CLASS MACHINE SHOP

Traits of a World-Class Machine Shop, a 30-Minute Shop Assessment, and Selection and Training of Employees.

Defining a world-class machine shop begins with *Chapter 1: Traits of a World-Class Machine Shop*—this chapter attempts to define world-class goals and prepare one for a shop visit. *Chapter 2: A 30-Minute Shop Assessment* discusses a quasi shop visit. This visit describes a walk-through of a shop with descriptive examples from actual shop tours. Questions that I ask during the visit are covered along with discussion of what the answers tell me. *Chapter 3: Selection and Training of Employees* is considered of high significance toward meeting world-class goals and is especially important to product quality.

CHAPTER 1

TRAITS OF A WORLD-CLASS MACHINE SHOP

Being a successful supplier requires more than average capabilities in personnel and equipment. Here is an attempt to define the traits and features of an ideal world-class machine shop. Many of the traits can certainly apply to any fabrication factory, but the core thought here is to define the characteristics of a machine shop that machines titanium alloys.

Key Characteristics or Traits of that Ideal World-Class Shop:

Employee Selection and Training

People are carefully selected, fairly treated, and well trained. They are customer oriented. These characteristics apply to the entire workforce. All employees (especially managers) are selected for their "fit" with the rest of the team. The company must be willing to take risks and let the staff make mistakes. Everyone has a say in the operation. A key philosophy is to be open to change.

Several companies have discovered that the quality of all personnel, not just machine operators, must be high to meet customer needs. They therefore set up training programs that included both machine operators and staff. This training is sometimes linked to a local community college, then the link becomes a conduit for recruiting. The courses include NC (Numerical Control) programming for milling in

Chapter 1

CATIA (Computer Aided Three-dimensional Interactive Application) or in Unigraphics, Lean Manufacturing, and Process Control. In top companies, the in-house training equipment is exchanged on a regular basis to stay abreast of the latest technology. This equipment and software exchange should take place about every three years. Training in all facets of fabrication and management has to be ongoing and not just a one-time happening. Grouping managers and operators together for training helps to break down communications barriers and labor/management boundaries.

Workplace Must be Safe and Clean

Is the factory safe and secure from outsiders, clean and organized with good employee comfort (lights, temperature, ergonomics, and free from harmful contaminants)? The entire shop needs to be clean with carefully marked off areas for everything. Titanium burns and the fires are difficult to extinguish, so special care is needed when making and storing chips.

Flexibility for Fast Creation of Work Cells

The ideal shop will have solid concrete floors at least 6 in. (152 mm) thick, and provisions for quickly moving light equipment to form efficient work cells as required. This flexibility can be most efficient in a shop with a thick floor—machines that do not need foundations, and machines that do not need to be bolted to the floor.

When I designed a factory a few years back, we hung all electrical lines with receptacles just above head level. Every machine in the shop used 440-volt 3-phase current and every machine and outlet were phased alike so the machine motors would rotate in the proper direction regardless of where plugged in. All air lines were routed with only one kind of fitting—tees. No couplings or elbows were used. This allowed us to add extra airdrops without major expense. Using all tees is not an extra cost because elbows, tees, and couplings all cost about the same; the initial labor for installing them is about the same as for other fittings.

Now with solid floors and useful grids of air and electrical outlets, equipment can be re-arranged into useful work cells in a matter of minutes without the time and expense of bringing in outside contractors such as electricians and pipe fitters. See *Chapter 20: Work Cells,* for further discussion.

Machining Quality

Part quality applies not only to the fabrication process, but extends also to handling, storage, and protection from contamination. Proper equipment (machines and fixtures), cutter checking at the machine with laser equipment or with fail-safe pins, and careful attention by all employees will show up in the records of cost of scrap and rework. In great shops these costs are less than 1% of total costs. One shop that I visit on a regular basis has used laser cutter checkers on each machine for over eight years and their scrap and rework costs averaged 0.3%. This shop also has a great record of on-time delivery. If you don't make scrap, it is easy to make your delivery commitments.

Well Maintained Equipment

Equipment should be well groomed, maintained, and calibrated on a regular basis. A preventative maintenance program that involves the machine operator is also required. Policies need to be in place to minimize spindle and tool holder run-out to within 0.001 in. (0.025mm) TIR (Total Indicator Reading) to meet the drawing requirements of most airframe manufacturing. This also improves spindle life and gives longer cutter life. I look for employee involvement in a TPM (Total Preventive Maintenance) program.

Efficient Operations and Processes

Are the operations and processes lean and efficient? Here I look for leanness in the non-value-added portions of the process as well as in the value-added operations. Examples:
- Capable and well-maintained equipment with a strong preventive maintenance program and key spare parts on hand or near by.

Chapter 1

- Efficient fixturing that makes use of a minimum number of fixtures that are properly rigid to minimize chatter and are easy to load and clamp.
- Part handling between operations that is fast and safe.

I urge you to also look for inefficient operations. For example, exceeding the quality contracted for by over-polishing is an inefficient practice.

On-Time Delivery

Is the delivery portion of bids well thought out and based on realistic quality performance? Of course, bad quality equals missed delivery dates.

Continuous Improvement

Lean activities are ongoing. There is a policy in place that limits the rate of growth. Several companies I know have allowed too much rapid growth to maintain proper quality. The influx of fresh people is just too much to handle and train properly. By too much growth, I mean greater than 50% in a business year. One company that I worked for had a strict policy that limited growth to a maximum of 30% per year to help insure that ambition would not overtake the ability to add properly-trained employees. What is your policy?

Does the company have a research and development budget? Technology is moving at a high pace and one needs to stay competitive. The amount spent on research depends on the characteristics of your products, but the answer of 0% is too low. Don't go too high in research spending either.

The ideal machine shop today is performing some internal research such as trying out new cutters and re-fixturing, but they are also doing some benchmarking with their competition, and they are learning to share technology:

Lean may not be enough—races do not have to be won by a lap or even 50 ft. (15 m). Races are often won by part of an inch or by a millimeter. I recently saw an order for parts, that cost $3000 each to machine decided by $1.84 per part. If most of your competitors are working hard on lean studies, what will you do to give your company some edge over the competition, something that will help your survival and profits?

It appears that lean studies are absolutely necessary and are the core of great improvements these days. But a few companies that formed a symposium have taken progress a step further. The most effective lean improvements that I have seen have happened within this symposium. Here, several competing companies actually conduct a round robin of lean studies. Each company, in turn, hosts a study with representatives from the other companies working and sharing together. The host company may also bring in technology experts and speakers; everyone shares further as they socialize and break bread together. Every three or four months a different company hosts a get together.

The members of the symposium have not become identical. They are of different sizes and selectively adopt or develop shared ideas that best fit their business and products. Here are some of the areas of progress:

- Designing, making and re-sharpening of your own cutters.
- Optimum use of cutter coatings for specific milling of titanium alloys.
- Use of fewer fixtures—the challenge is one fixture—or even better, universal fixtures.
- Cutter feeds and speeds that match or top the best that I have seen in the industry.
- NC Programming techniques and standardizations that are setting trends in the industry.
- Heading off large expenditures of capital equipment.
- A sharing of ideas on reducing machine down time.
- Group pressure on machinery builders in this country and abroad to improve machine designs; not just waiting to see what equipment manu facturers will come up with.

Chapter 1

We have started a second symposium with seven other companies, and after one year, great results are also happening in this new group.

Another area that is under consideration is to team up and share technology with members of the supply chain that tie to these machining houses such as: chemical houses, thermal processors, inspection services.

Are You Ready for Expansion?

Do you have a plan for your expansion as demand for product increases? The growth can take place either within present facilities, on present grounds, or down the street. I look for indications that a company is looking ahead and does have room for expansion.

Cutters

Cutters are a large expense and a potential work stoppage when not available. Here one needs a great process that assures a timely supply of cutters to the operators. The process or plan for cutters must be good, and allow for re-sharpening and re-coating when necessary. But most important of all, are you properly aggressive with your use of cutters?

Competitive Prices

The collective effect of the above characteristics or traits is sound competitive bids.

How Do People See Your Shop?

A shop takes on and has a personality. Some examples of shop personalities that I have observed: Pigpen (character from Charlie Brown comics), Miss Prim & Proper, Mr. Stud, Joe Casual, Mr. Business.

Examples and Discussion of World-Class Elements in Shops

The most efficient and lean shop that I know of is a small broaching company in Michigan. In this small shop, one lady tends the entrance door, answers the phone, and watches several children in a day-care room at her side. The shipping paperwork includes the billing for the job, so separate billing operations are not necessary. Accounting and payroll are covered by an outside person who comes in about two hours a week (longer at the end of each quarter to work taxes.) The owner is the marketer, fixture designer, and principle equipment maintainer. There is one toolmaker, and the rest of the employees run the twelve broaching machines.

The cleanest shop is in Sacramento, California. A factory of about 50 employees—so clean that I want to clean my shoes before entering.

The most flexible shops are two that move equipment by the machine operators on a solid floor with a pre-wired grid of power drops and a grid of airdrops. A fresh work cell of rearranged equipment can be set up and operational within one hour.

The most aggressive milling of titanium alloys that I have seen is by a California shop who took our research data of finish cutting with solid carbide at 600 sfm (surface feet per minute) (183 meters per minute) and 0.005 in. (0.127 mm) chip loads on up to 800 sfm (244 meters per minute). The norm had been 120 sfm (37 meters per minute). They are also doing research on coatings and new cutter geometries and have near-term goals of exceeding 1000 sfm (305 meters per minute) on finish cuts in titanium alloys.

The lowest overhead costs are in a Cincinnati company who own all their equipment free and clear. They run their mills at slower than normal feeds and speeds for extended cutter life, and they use only two operators per shift for about 14 machines. Their wrap rate is about 40% lower than most of their competition. Their risk, of course, is quality because of antiquated equipment.

My point is that there are many ways of attacking the task of becoming efficient and World-class. Each shop must determine what will work best for you—what fits your shop culture, and your circumstances?

Chapter 2

A 30-MINUTE SHOP ASSESSMENT

I am often asked to make an assessment of a machine shop, or questioned, "What do you look for in a shop?" A thorough assessment would usually take several days and involve extensive review of process plans, operator buy-offs, and investigation of internal operating procedures, but sometimes one doesn't have several days. So what can be accomplished in a short pass through a shop?

This chapter outlines a 30-minute walk through of a machine shop, and is an update of my paper (*A 30-Minute Shop Assessment*, published by SME.) Specific questions that I ask during a shop visit are listed. Unasked questions and thoughts that go through my mind are also mentioned. The reasons for these questions and the ramifications of the answers are discussed with examples. The normal intention of this brief walk-through of a company is to collect notes and to develop a mental picture of the factory that I can draw on later. I am often asked about a given factory's capability to produce certain parts. My concerns are about part size, right equipment for the job, well maintained equipment with spindle run out within 0.001 in. (0.0254 mm), expected part accuracy, anticipated quality, accurate bidding approach, lean activities, and a supplier's ability to expand to higher work volumes should the need arise.

Observations and examples of various shop and equipment condition are also discussed, but most importantly, the people are observed. A **Check List for a Shop**

Chapter 2

Assessment is included at the end of the chapter.

The Shop Tour

Initial Greetings, Safety Instructions, and Security

Last summer, I walked into a machine shop back east with no challenge whatever. The person who had invited me to visit had forgotten that I was coming that day. The lobby was unattended and doors were unlocked. I was able to wander all over the shop and observe parts being machined with no discussion with anyone.

I was uneasy about this experience, and the following thoughts and questions went through my mind:
- What are their rules on safety glasses and side shields?
- Are there any danger spots in the shop: chemical danger, mechanical danger, construction, or…?
- Are my parts safe from theft? We have a case or two each year where material is stolen (likely for sale as scrap.)
- We also have proprietary agreements in-place with the majority of our suppliers. We certainly do not want strangers off the street to have access to these papers.

General Cleanliness and Orderliness of Shop

I look for a shop to be reasonably clean and uncluttered. Orderliness usually means marked-off areas with labels and signs. Things that worry me are too many tubs of chips and excess parts and pallets scattered throughout the shop. People have to walk around them and move them to get at whatever it is they are working on. Safety is an issue too if the chips happen to be titanium, which is a fire hazard. On the other hand, I remember three shops that were so clean that I felt compelled to wipe my shoes before I entered.

Condition of Equipment

Does the equipment look "well groomed?" What is the "up time" of your machines? I also ask about frequency of preventative maintenance and availability of spare parts that are likely to fail often. How often is the equipment calibrated? How about calibration after a crash or spindle change? What spindle run-out is tolerated?

Here I look for no more than 0.001 in. (0.025 mm) of run-out. This run out is often referred to as TIR (Total Indicator Reading). I did find one shop that tolerated 0.007 in. (178 mm) of run-out and called maintenance if run-out was above that figure. In this instance, we were experiencing a reject tag on almost every part. The problem with high spindle run-out is poor part quality and low tool life. I have seen milling cutters with the teeth completely worn out on one side of the cutter and brand new teeth on the other side. My math says that these cutters would last at least twice as long with zero run-out, or, as an option the chip load per tooth could be doubled.

What is your spindle draw bar pull setting? I like to see it about 8000 pounds (3629 kg) to prevent any cutter pull out. 2000 pounds (907 kg) is not enough for aggressive titanium alloy rough milling. I am referring to rather large machines here with 50 taper tool holders.

Tool Holders

I always ask what run-out is tolerated on tool holders? Once again my recommendation is no greater than 0.001 in. (0.025 mm) TIR, otherwise tool wear will be uneven, as mentioned earlier. My preference is to use heat shrink tool holders because they tend to run true, are readily balanced, and are reasonably priced.

The Facility and Flexibility

How do they move heavy objects—overhead cranes, or…? Are the floors thick enough to allow small machines to be rapidly re-located without needing special foundations? This flexibility allows easy formation of new work cells for better efficiency. I look for a minimum of 6 in. (152 mm) thick reinforced concrete floors; 8 in. (203 mm) is better.

Do they have room to expand their machining operations either inside the existing buildings or on nearby property?

Chapter 2

Process Paperwork

I look briefly at the process routing sheets for two main items:
- Recording of serial/lot numbers (traceability information) near the beginning of the shop orders.
- A place for the operator to stamp-off each operation. I also look to see if the operator is stamping each operation when it is completed, not a massive stamp-off from memory just before the job ships out.

Shop Efficiency

Are staffed machines cutting chips? In one Boeing machine shop, shortly before its demise, our delay studies showed us cutting chips 40% of the time when operators were clocked in to the job. Most of the non-productive time, which added up to 60% was due to delays for setup, no operator present, and unplanned machine down time. Many of our good suppliers are cutting chips about 60% of the time when operators are clocked in to the job. My favorite tool for measuring machining efficiency is to perform a delay study, but not as part of the 30-minute walk through. I will discuss this later in Chapter 19.

Are employees clocked in to the given part or job? Some shops do not record actual costs of given parts. I wonder how they know what jobs to work on for improvement? Are they making money on given parts? How do they answer the question of "When will the part ship?" How sound are their bids on new work if they do not have a specific history of costs? Are lean activities happening? A lack of improvement will adversely affect future cost trends.

Employee Training

I like to see employee selection based on how they seem to "fit in" with the rest of the team. Training needs to be ongoing and continuous in these fast paced days; I consider it a plus when management and fabrication folks are trained in combined sessions.

Assessment

Some questions I ask are: What is your source of new employees? Are you tied to a local college? How about in-house training? What is your turnover rate in the areas of machine operator and NC programmer?

Attitude of Management

Is there a positive attitude—a sense of teamwork—a zest for quality and on-time performance at all work levels?

Attitude of Workforce

I look for employee involvement here. The answer is discerned from observations and discussion with employees during my walk through.

Fixturing, Fail-safe Pins, and Setup

One of the first things I look for here is the use of fail-safe pins or equivalent mechanisms for assuring that the correct tool and cutter set lengths are used. A fail-safe pin is placed in a socket in the part holding fixture for milling. The pin usually has a frangible or easily broken off top section, and the NC program has routines that take a rotating cutter around the pin after each cutter change. Running around the pin checks that the cutter is not too large in diameter. The cutter is passed up along the pin at a 45-degree angle to make sure the cutter radius is not too small. The cutter is also passed across the top of the pin to make sure the cutter does not stick out too far. An undersize cutter can pass this test and would not remove excess stock from the part—hence the term fail-safe.

Before we introduced fail-safe pins in our Kent machine shop our scrap and rework cost was nearly 20%. After introduction of fail safe pins, the cost of scrap and rework dropped to under 2%, and schedule performance improved accordingly.

I look for a clearance of 0.002–0.003 in. (0.051–0.076 mm) between edge of cutter and fail-safe pin, but found a shop with 0.010 in. (0.254 mm) of clearance. This supplier machine shop needed the wide clearance to accommodate dilapidated equipment that had excessive spindle and tool run-out. Spindle run-out was as great as 0.007 in (0.178 mm) and tool holder run-out was as great as 0.009 in. (0.229 mm). This adds up to a potential combined run-out of 0.026 in. (0.66 mm). The wide gap between cutter and fail safe pin was revealed to us (the customer) when a part was scrapped; it was cut too thin by use of an oversized cutter that actually passed the fail-safe pin test.

Are fixtures easy to load and unload? I had a seasoned toolmaker from Europe for a few years that designed and built great tooling for us. The clamping levers were always out front were the operator could easily get at them, not on the backside or underneath where you couldn't readily reach them. His stamping dies always chopped up the strips as they passed through the press, or he had a roller that rolled up the strip stock as it exited the die.

In the past couple of years I have seen many shops reducing the number of fixtures for milling. In one example, three fixtures were reduced to one rotisserie fixture. With fresh programming, the total time went from 20 hours to less than 2 hours.

So I look at the fixtures to see if the shop being assessed is creative with fixtures and uses efficient tooling.

NC Programming

What NC programming language is used? The answer is important if we need to move a job between shops or even within this shop. In one supplier shop, half of the machines have unique controllers and the other half have standard controllers; there is no software for reposting NC programs when it is necessary to move jobs between machines. In this shop, reprogramming is necessary to move a part between controller types or to bring a part in from an outside shop. Non-standard controllers can result in high costs, in a loss of days, additional programming effort, the added costs of doing a machine tryout part, and a new first article inspection each time the job is re-programmed. Most of our suppliers use CATIA or Unigraphics programming today with Vericut or similar electronic checking to look for potential crashes. Some of the top shops also add CATIA models of fixturing, spindle, and machine features

Assessment

into the models for Vericut checking.

Can the programmers work from supplier's CATIA models? I visited a shop in Los Angeles who has six programmers, but they are programming in five different languages. Each NC programmer programs in the language he or she is comfortable with. How do they back each other up when vacations happen or when a programmer quits? What happens with overload when they hire contract programmers? What is the turnover rate among programmers, and are contract programmers ever used? The importance of standardizing on one programming language is the efficiency of using the latest high speed milling techniques and cutters with consistency. We want our costs and flow time to be as low as possible.

Milling Feeds and Speeds

For titanium alloy rough milling I look for milling to be 50–60 sfm (15–18 meters per minute) when using cobalt cutters (cobalt cutters are nearly always made from M42 cutter material which contains 8% cobalt), and for finish milling. I would traditionally expect 100–120 sfm (30–37 meters per minute) for carbide cutters (either solid carbide or carbide inserted cutters.) Solid carbide cutters are made of micro-grain or sub-micro-grain carbide which means the grains measure no greater than 0.5 microns. Today, efficient shops are using 400–800 sfm (122-244 meters per minute) for finish milling using carbide cutters with TiAlN or equivalent coatings. Many shops use wave cutters for roughing (or similar tools such as corncob cutters.) One shop is finish milling at speeds up to 800 sfm (244 meters per minute.) I look for all cutters for titanium alloys to have positive rake and to be very sharp and for heavy use of flood coolant during all milling.

Cutting Tools

Here I want to know who resharpens their cutters. Are they using coatings properly? What is the timing (wear factor) for replacing cutters? Do they recoat after regrinding? One shop I know hires an outside firm for the task of putting cutters in

Chapter 2

tool holders and setting the tool length; the hired people are over 50 miles down the road. The long distance results in two-days turn around vs. a couple of hours if the tool company were across the street. The first cutter I checked for was two weeks late getting replenished.

Assessment

Check List: Shop Assessment

Safety glasses, side shields, face shields, ear protection? _____

Building security? _____

Cleanliness and orderliness? _____

Titanium chips safely stored? _____

Equipment well groomed? _____

Well marked off areas? _____

Milling feeds and speeds? _____

How often is equipment calibrated? _____

Spindle run-out within 0.001 in. (0.0254 mm)? _____

Drawbar pull set to a minimum of 4000 pounds (1814 kg)? _____

Tool holder run-out within 0.001 in. (0.0254 mm)? _____

Cranes? _____

Floor thickness 6-8 in. (152–203 mm) reinforced? _____

Room for expansion? _____

Lot/Serial Number tracking on planning? _____

Operator stamps operations upon step completion? _____

Estimate of shop efficiency? _____

Do employees clock in to the job? _____

Lean activities happening? _____

Employee training? _____

Chapter 2

Attitude, teamwork, zest for quality? _____

Employees involved in decisions? _____

Fail-safe pins with 0.002 in. (0.051 mm) gap to cutter?_____

Rigid fixtures? _____

Fixtures easy to load? _____

NC Programming language/s? _____

Work from CATIA models? _____

Programmer turnover rate? _____

Sfm for cobalt and carbide cutters? _____

Who resharpens cutters? _____

Are cutters coated and re-coated? _____

CHAPTER 3

SELECTION AND TRAINING OF EMPLOYEES

Have fun with your job—we spend a lot of our life at work. I am convinced that my job is good for my health, and the fact that I enjoy my work probably makes the job even healthier. I am still working full time at age 70 and my medical and heart doctor agree that working is good for me and good for my health.

Perhaps we should foster a work environment, atmosphere, and attitude in our factories that extends the notion of every employee having fun with their job.

These are tough times for machine shops. There is strong competition from other shops and from other countries. Progress in technology, improvements in management techniques, and getting employees involved are all progressing at a fast pace. Many machine shops are not as up-to-date as they need to be to stay ahead of their competition. Training in all facets of the operation can help solve this need.

The purpose of this chapter is to suggest areas for training that would likely be beneficial to small- and medium-sized machine shops, and to provide some ideas and examples for each topic. This chapter suggests areas for training in machine shops. The underlying **objective is for your shop to become ever more competitive**. Some ideas and examples are included for each topic to help point out the importance and expectations for each area of discussion. This writing is certainly not an employee-training course, but is an attempt at providing a solid outline for training.

Areas for training should probably include:

Chapter 3

- Leadership and employee involvement
- Management selection and attitudes
- Employee treatment and involvement
- Manufacturing research and development
- Cutting tools
- Fixturing and job setup
- High speed milling
- Part warpage—prevention and straightening
- Coolant ideas
- PMT – preventive maintenance training

Leadership and Employee Involvement

In our definition of world-class employees, we are looking for people who were carefully selected, fairly treated, and well trained. They are customer oriented, and fit well with the rest of the team. Training is part of the equation of arriving at such a work force.

Management Selection and Attitudes

Here one might ask for help from other shop managers. You might want to interview or bring in a manager or two from companies that seem well run to ask them to share their ideas on people selection and on training. One president of a machine shop that I interviewed makes a strong effort to hire and place staff members based on his assessment of how well they will fit-in with the balance of the team. The focus here is on "fit" and "teamwork," and not necessarily on the greatest intellect or highest-skilled person available. This philosophy seems to produce great results for this particular company, and helps to minimize turnover.

Training of management on how to treat employees as individuals and in being particularly tuned in to their requests and needs is the likely theme here. My strongest recommendation is that you teach great listening skills.

Training

Employee Treatment

Several companies discovered that the qualities of all personnel, not just machine operators, were needed to meet customer needs. Therefore, they set up training programs that included both machine operators and staff in the same sessions. Training managers and operators together is a great practice that breaks down communications barriers and reduces boundaries between labor and management.

Training is sometimes linked to a local community college; then the link becomes a conduit for recruiting. The courses include NC programming in CATIA or Unigraphics, Lean Manufacturing, and Process Control. In some companies, the in-house NC training equipment and machines are exchanged on a regular basis; say every three years, to stay even with the latest technology. Training has to be ongoing and not just a one-time happening to keep pace with rapid advances in technology and memory decay.

NC Programming

Training here needs to include instruction on specific brands of software. What are our company needs and goals, and how extensive should we perform electronic checks on programs? Should we re-verify and check even after minor revisions to NC programs? The answer better be yes. One company that I work with insists that full electronic checks be made even following the tiniest of NC program corrections.

Efficient Programming

Software for programming: first the part is created in CATIA, and then converted to Vericut to get the g-codes. Then Vericut verifies each program. This eliminates a lot of programming error. The program is expected to be error free when it goes to the shop floor.

As an example of the precision of the programming and equipment, in one case, large holes are bored from either side instead of the traditional through boring. This is possible because of the high machine accuracy.

Chapter 3

NC programs emphasize heavier than usual roughing cuts so finish milling can take less time.

In the most successful shops, on all NC programs, even slight changes are subjected to a full electronic confirmation using software like Vericut, and in great shops the NC programmers add the elements of machine spindles and part fixturing to the part computer models to look for interferences and crashes beyond just the tip of the cutter. It appears that industry is close to eliminating the need to machine a first part to try out or confirm new NC programs.

I continue to see programming-caused inefficiencies in every shop. Programmers need to be challenged to wring out that extra drop of moisture from the rock. Here is a short checklist:

Find the shortest cutter paths between cuts

Cut in both directions (when this makes sense.) We stopped conventional milling about 40 years ago and generally only use climb milling, but the other day I saw a shop milling in both directions (climb milling then conventional milling.) With good equipment and the right conditions they had eliminated the return air cut.

Acceleration/deceleration in corners—even more important as we increase feed rates.

Rapid travel while air cutting—many media have some of this inefficiency.

Start feed rates close to the cut. A good test of an efficient program is when the cutter doesn't start slow feed until it is about 0.005 in. (0.127 mm) from the part and the operator begins to panic and starts to reach for the red stop button. Of course, this boldness can only be successful if one is sure where the "surface terrain" of the metal starts.

How do I keep employee turnover low? What are proper plans for bringing in contract programmers?

Lean Studies

Lean studies are the rage today and for good reason; they cut costs and flow

time and can even improve quality. There are many training opportunities. Besides the fundamental lean studies you should include training on:
- Maintaining processing paperwork
- The importance of maintaining accurate cost-of-job records
- Preventive maintenance that makes use of the operator's eyes and ears. This training is called TPM or total preventive maintenance, and the training takes place at each person's machine and associated equipment.

Sometimes your key customers will come in and train you in many of the above areas at no cost to you because they are anxious to lower the cost of their purchases from you. (See Group VI below for further discussion of lean.)

Summary

Suggested training actions are:

- Set an atmosphere for having fun on the job.
- Train everyone in the factory, and group management and operators together for much of this training.
- Train in every facet of the job.
- Make training an ongoing happening, focus on lean studies.
- Train for machine maintenance (TPM.)
- Link up with local community colleges when this makes sense.
- Take advantage of free training from major customers.

GROUP II: EQUIPMENT

Work Holding, Fixturing, and Machines.

How does one equip a world-class machine shop? Here the term equipment refers to work holding and fixturing and also to machines. *Chapter 4: Work Holding and Fixturing* describes creating fixtures that are efficient to use yet inexpensive to build and maintain are the challenges here. *Chapter 5: Machines,* gets at the heart of milling parts. What is the right equipment, and how important is it to have the right equipment?

CHAPTER 4

WORK HOLDING AND FIXTURING

When I make presentations on fixturing and work holding, I try to accomplish two objectives:
1. Challenge machining people to be impressively creative in the use of fixtures and attendant processes for large part milling.
2. Show ways to improve quality and make major improvements in flow-time, and costs of machined parts.

Fixture Concepts that Improve Efficiency

Fixture efficiency can be improved by using fewer fixtures, rotating the fixture, and making fixtures more rigid to reduce chatter. Reducing the chatter allows more aggressive milling. The use of dovetail fixtures on small parts allows milling on five sides of a part in one fixture. Further efficiencies can be achieved with universal fixtures that handle several part numbers. The use of tooling tabs can sometimes eliminate the need for special fixtures.

<u>Fewer Fixtures</u>

A major contributor to success by several machining suppliers is to use fewer fixtures. Fewer fixtures mean less non-recurring fixture costs, greater part accuracy, less setup time, less work in process, and less storage cost. Changeover between multiple fixtures uses up valuable time and results in introducing more chances for part error through mislocation of the part or fixture, hence lower part quality when using many fixtures.

Around a year ago a group of seven machine shops that supply machining services to our company met and exchanged technical ideas on fixturing. They also posed a one-year challenge to one another to see who could return with the greatest fixturing ideas. The previous approaches to milling often ended up with six fixtures per part such as:

1. Roughing side A
2. Roughing side B
3. Intermediate milling of side A
4. Intermediate milling of side B
5. Finish milling of side A
6. Finish milling of side B

Out of this challenge came many great improvements, with most parts now needing two or less fixtures. Perhaps the greatest improvements came with the use of rotary fixtures.

Rotary Fixtures

Consider the introduction of rotary fixtures. Here one milling fixture can take the place of several and still maintain the advantages of flip-flopping the part to minimize warpage. Manually-driven and program-driven rotisseries along with the idea of making these fixtures adaptable to families of similar parts can even further reduce fixture costs and changeover times.

Rotary fixtures allow the tool to access more of the part's area in a single fixture. Most of these fixtures are rotating fixtures that essentially add one or two additional axes to the milling machine. Most milling machine controllers have the capacity to drive additional axes such as the motor of an indexing fixture along with the slave motor driving the far end of the part. Picture the part mounted on a rotisserie, as in a home grill. The rotisserie-type fixture allows more of the part's surface to be presented to the cutting tool. One trick to the success of rotisserie fixtures is to positively drive both ends of the fixture. By driving both ends of the mounted part, torque and twist are minimized in the part.

More than one rotisserie can be used on a multi-spindle mill. In one case, three rotary fixtures replaced nine flat fixtures (see Figure 4-1). This reduced setup time by 67%, and was done on a 3-spindle five-axis gantry machine—so there is actually one rotary fixture per spindle. If rotisseries are long, a steady rest feature is added near mid-part, much as is done with a lathe operation. All the principles we have learned about using lathes on long parts over the years apply to the use of rotary fixturing.

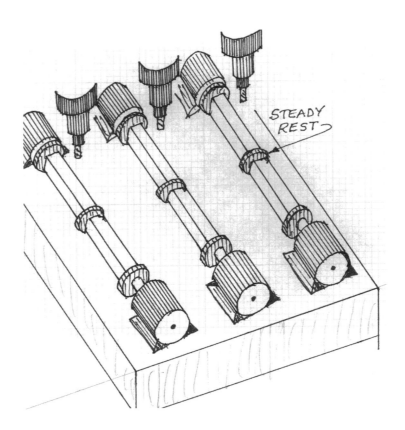

Figure 4-1: Rotisserie Fixtures

Chapter 4

These same rotisserie fixtures are designed with great rigidity to allow high cutting rates and to minimize chatter.

On a single spindle machine, my favorite fixture is a single rotary fixture I saw for a part shaped like a hockey stick about 7 ft. (2.1 m) long (see Figure 4-2). The old approach used three fixtures with about six hours of setup time—two hours per fixture. The old run time was about 14 hours for a total of 20 hours. Today this part has a floor-to-floor time of less than two hours including setup time. NC programs were also revised to reduce the run time, but going from 20 hours to less than two hours is not bad, and I doubt if any other shop will bid this job away from the existing company.

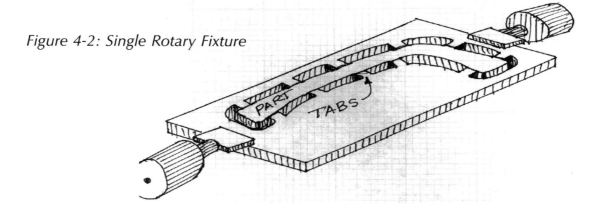

Figure 4-2: Single Rotary Fixture

For a family of parts that vary only in length, you could use a common rotisserie for several part numbers. Here one adjusts the slave end of the rotisserie fixture for part length for each part in the family.

Making Fixtures More Rigid

Fixture improvement, especially when rigidity is improved, creates an opportunity to introduce high-efficiency rough cutting and high-speed finish milling that will further reduce costs and improve quality (this is needed for milling titanium alloys.) When chatter is encountered while milling, the speed of removing chips is limited in order to maintain proper surface finish and good cutter life. Chatter, the

Fixturing

main culprit, when milling titanium alloys is vibration within the part, fixture, and tombstone (see more on chatter in *Chapter 6: Rough Milling of Titanium Alloys*.) Use of rigid fixtures and dampening of fixture/part/tombstone is usually the solution to chatter when cutting titanium alloys.

Dovetail and Bolt-Ons for Small Parts

For small parts, say parts the size of a large lunchbox; we often cut a standardized dovetail in one side of the raw material (see Figure 4-3). This allows the part to be quickly anchored into a universal dovetail-holding fixture on the mill. Usually we can then mill out the entire part because we can machine five sides of our rectangular stock in this one setup.

Instead of dovetailing, we sometimes drill and tap two or three holes in one side of our stock and bolt the material onto a standard fixture. We get the same advantages as with a dovetail fixture; we can mill five sides of the part in one setup, much as described above.

Pogo Tools

A pogo tool (sometimes called a bed-of-nails) is a universal fixture for trimming

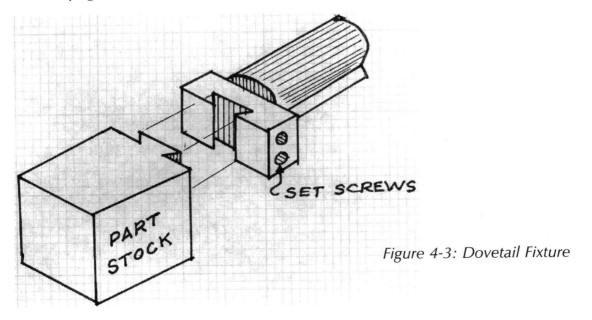

Figure 4-3: Dovetail Fixture

Chapter 3

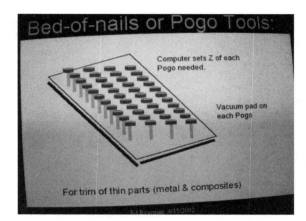

Figure 4-4: Pogo Tool

thin, flexible five-axis parts. It is used for both sheet metal and composite parts (see Figure 4-4). The center spacing of the pogo grid posts is about 5 in. (127 mm). Each post is set for height, the top vacuum pad swivels to match the part contour when the part is placed. A small part works fine because only the pogos under the part are activated. Edge part locators are used to position the part. There are several methods used for setting the height of the pogos. Our main fixture is used for trimming about 100 different parts.

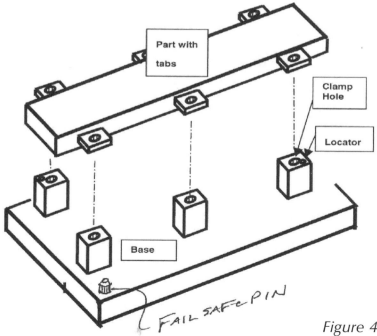

Figure 4-5: Sketch of Tooling Tab Use

Fixturing

Tooling Tabs

Some parts have a convenient flat surface to support and locate the part against a tooling plate during milling. An option is to have built-in tooling tabs for supporting and precisely locating the part in fixtures (see Figure 4-5). Built-in tooling tabs require drawing definition of excess material on the casting, forging, or raw plate for the tabs. These tabs are positioned around the periphery of the part. In cases where the part is rather flat, without much variation in Z height, milling flats on the top and bottom of the tooling tabs establish the plane of the part. The part then sets on standoffs from the tooling base. A hole is drilled in each tooling tab for clamping into the threaded hole in each standoff, and part location is established by using a round pin in one standoff and a diamond pin in a second standoff to mate with like holes in two of the tooling tabs. The same locating and clamping through holes in the tooling tabs are used when the part is flipped over for milling on the other side.

If there are no built-in tooling tabs, then one of the following approaches might work. A clamping groove or slots can be machined along the periphery of the part (see Figure 4-6). These grooves are usually located about midway up the edge of the part and allow standard clamps to hold the part down. The part can rest on its flat surface, but if there is no flat side, support pad spots are machined into both sides of the non-flat surfaces.

To prevent bowing of the part, clamps are always near stand-offs for the support pads. If the raw material does not have enough excess stock for tooling tabs, then pads are sometimes bonded or welded onto parts to locate and support the part dur-

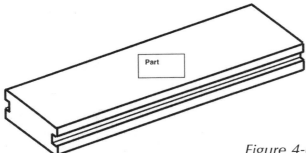

Figure 4-6: Sketch of Part with Clamping Grooves

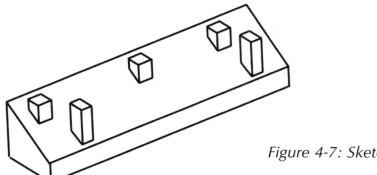

Figure 4-7: Sketch of Tabs, Bonded or Welded

ing milling (see Figure 4-7). Drawing permission is usually required for use of these tabs to make sure that bonding agents do not adversely affect the material of the part or, in the case of welded-on tabs, that the heat-affected zones are areas that will be completely removed in final milling. After bonding or welding, the standoffs, which are usually long enough to establish a flat plane, are faced off in planer fashion.

In all fixturing and clamping schemes, the question comes up, "What is the proper spacing between supports?" Many of the peripheral tooling pads are about 16 in. (406 mm) apart, but if your tool designers are looking ahead toward fastest possible chip removal, I suggest further engineering. As discussed earlier, chatter usually limits the RPM of roughing cuts in titanium alloys. Remember to work the fixture/part/tombstone to minimize chatter when milling titanium alloys.

Another consideration on strength of fixtures is to allow for future improvements. A couple of years from now some creative Manufacturing Engineer might stumble across a new cutter with a high helix angle and attempt more aggressive roughing cuts in titanium alloys. When I tried this a few years ago on some of our parts, I had big problems. Some of the fixtures were too weak and bent or tore loose from their hold-downs.

Another problem was that I pulled some cutter holders right out of the spindles. Here the drawbar pull was set too low (about 2000 pounds) (907 kg). I had them re-set to about 8000 pounds (3629 kg) and this problem went away. Perhaps fixtures should be strong enough to withstand 8000 pounds of pull from cutting. Then the efficiency folks could work their future magic. Here I am talking about large milling machines with 50 taper spindles for milling titanium alloy parts that are perhaps 2 meters long or larger.

Setup Time Considerations

Thoughts on saving costs by having efficient setup equipment are discussed below.

Pallet Equipped Machines

Pallets on many machine tools are set up to carry two work pieces. While one part is being mounted and set up, the other part is being machined. This makes setup time internal to run time and saves both machine and schedule times.

Big Machine Beds Allow Fixtures to Remain in Place

Some large gantry mills have such large beds that several fixtures can be left set up indefinitely even though they are only used a few times per year or month. Here setup time only costs the time to load and unload parts into the fixture. Except for the cost of space used by the large machine footprint, this is almost as efficient as having a machine dedicated to a single task.

Fast Fixture Orientation

Fixture setup is an area where some in industry are further advanced than others. There are at least two setup methods that beat the old way of tapping fixtures back and forth with a mallet until dial indicators show that the fixture location is good enough.

Tooling Balls and Valysis

One setup method that is fast is to use tooling balls and to "find the part or fixture" with probes. Here we use three-to-five tooling balls—placed on the part or fixture. This setup method requires the use of Valysis-type software, and requires a controller that can electronically re-skew the machining part planes. The probe data allows the machine to find the part much like a CMM inspection machine. On-machine probing and the use of tooling balls on parts and fixtures allows early part checking without leaving the mill. This method also facilitates efficient re-setup of part if needed. When I talk about probing on the mill, I am not talking about full-up CMM-like part inspection, but quick checks of a few key part features that indicate that the part is accurate and that it is ok to proceed to the next step. The machine controller or program is then re-set to conform to the new part location.

Replaceable Tooling Pads

A second method of quick fixture location is to have replaceable locating pads in a base fixture that are re-drilled for each succeeding setup. Here the machinist who is going to machine the part, drills the two locating holes. The upper portion of the fixture with the part is then placed and pinned to the freshly-drilled base plate and clamped or bolted into place. Setup time in both of these processes is a matter of about 15 minutes rather than an hour or more. Precision setups using previous techniques can take up to 6 or 8 hours.

The traditional processes of old sometimes required one or more breaks in setups, with intermediate trips to the CMM machines for checks to mark parts for hole corrections and for confirmation that it is okay to start the next step in the process. Besides costing a lot of time, re-setup of the part, back at the machine, is a potential source of dimensional error.

Fixture Materials

On most small or medium size parts, the expansion and contraction of fixture materials is not large enough to affect the specified part dimensions, but on large airframe parts made of titanium alloys, one must be cognizant of the dimensional compatibility of the fixture material with the material of the part being milled.

Fixturing

One can usually select low cost fixture materials that are temperature compatible with the part material. I often use commonly-available mild steel as the base material for fixtures that hold titanium alloy parts. The steel selected has a CTE (Coefficient of Thermal Expansion) of 6.6×10^{-6} in./in./°F (degrees Fahrenheit) (11.9×10^{-6} mm/mm/°C). The CTE of our commonly-used titanium alloy ti-6Al-4V-ELI-BA is 5.7×10^{-6} in./in./°F (11.0×10^{-6} mm/mm/°C). The ELI in the alloy description stands for Extra Low Interstitial and in simple terms this means that less oxygen is allowed in this alloy which gains material toughness but looses some strength in the trade-off. The BA at the end of the alloy description stands for Beta-Annealed (heat treated.) The mild steel is also easy to fabricate and weld, and is one of the few materials that do not have to be stress relieved after welding.

Chapter 5

MACHINES

Discussions in this chapter are primarily about large machines for milling titanium alloy parts above 4 ft. (1.2 m) in length. We purchase most fasteners and turned parts for aircraft, so equipment such as screw machines, lathes, and grinders are not discussed in this book. Besides, most of my experience is with milling.

Long Range Plans/Strategy/Mission

Your company needs to have long-range plans. These plans will guide you in making capital investments. Here are specific examples of what three different companies did to solve the same milling need on a given aircraft weldment. The part envelope was about 10 x 4 x 4 ft. high (3 x 1.2 x 1.2 m) and required five-axes equipment. The examples of what three companies did to machine these parts are:

- Company A purchased a new mill with many bells and whistles for about $6 M and installed the machine in a lowlands area where pilings and foundation cost an additional $3 M for a total cost of $9 M.
- Company B found a more modest machine that cost $2 M installed on a special foundation, but in a location where pilings were not required.
- Company C modified a couple of existing horizontal boring mills and did not have to re-do foundations. They converted horizontal boring mills from three-axis to five-axis by:

1. Changing to a nutating head that added the A and B axes
2. Changing to a five-axis controller
3. Extending the Z-axis along the floor to allow the part to move up to 6 ft. (1.8 m) away from the head in the axial direction. Total cost to company C was $0.5 M per machine.

The outcome of these investments after ten years is:
- Shop A with the $9 M total investment was disbanded for high costs and no longer exists.
- Shop B with the $2 M investment had to buy a second $2 M machine for increased capacity and is still milling the 10 ft. (3 m) parts. Total investment was $4 M.
- Shop C is also still milling the 10 ft. parts after re-working a 3rd boring mill with a total investment of $1.5 M.

The point is that many solutions exist for a given problem and we can certainly learn from others' experiences. Do your research and homework, it just might save you a lot of money. The company who spent the $9 M got caught up in a trap that you need to avoid if possible. The trap was that the persons in the machining company, who were entrusted with the task of specifying the machine to be purchased, wrote the specifications so tight and unyielding that there was only one machine and company that could meet the needs. So competition between machine suppliers did not happen.

Financial Considerations (New vs. Used)

Every factory has its own culture and needs to calculate what is best for them and develop their own strategy for machine procurement. Here are examples of some companies' strategies that are being practiced.

One company that I work with buys only used machines. They also buy extra machines for spare parts. Their policy is to minimize capital investment in order to achieve lower overhead and bidding wrap-rates than the competition. This company is trading capital costs of machines for less new technology, but that is their long-range plan.

Another company buys new and re-built machines from only one company—one brand name. Their agreement with the machinery builder means that they do not need to start paying for the machines until one or two years after the machines are installed. Further costs are saved in the stocking of spare maintenance parts and in the cost of training for machine maintenance. There is no interest accumulated and the rates are below 7%. This gives the company a year or two to get work loaded on the machines to establish income from which payments can be made. The obvious risk here is in being able to get new work for the machines, but then, we all have that challenge.

Other companies buy a variety of machines (new and old) and competitively bid each order. This allows them to acquire new technology at competitive prices, but they do have to start payments for the machines early.

There are many other strategies for machinery buying. What is best for you?

Families of Machines

Except for the very small company, most companies have families of machines. The obvious reasons for this are:
- Ease of moving operators between machines
- Common use of NC programs and fixturing between machines
- Reduced cost of having an inventory of key spare repair parts
- Less training for machine service
- Identical preventive maintenance programs
- Cost breaks for buying machines in large quantities.

An extension of this policy is the ability to offload to companies that also have the same brand of machines (provided your brand is not unique and one of a kind.)

Machine Location and Future Re-location

If your newly acquired machine requires a special foundation, then it is certainly important to put some thought into where to place the machine relative to other machines and process operations. Here are two examples of placement that did not work well:

1. A machine with a special foundation is located about 20 ft. (6 m) from a busy railroad. Fifty-four trains pass this machine each day. Unfortunately, the

milling patterns on parts become a permanent record of the trains passing. The variations in milling can actually cause part rejections, or at best, one pays for additional sanding and benching to save the parts. This factory is abandoning the railroad property and is paying for a new foundation in a fresh location away from the railroad.

2. A factory is built on a nice flat old flood plain; it is 60 ft. (18 m) down to hard pan soil. Special foundations under heavy machines require pilings that are over 60 ft. deep. The cost of the foundation for one large machine with a footprint of 100 ft. (30 m) in length was about $3 M.

Part Requirements

Here one determines the size of the largest part anticipated for the machine. Remember to include the size of the fixture as part of the total envelope. Last year, one of our suppliers was anxiously awaiting the startup of three new five-axis machines that were purchased for a specific family of parts. Our production needs had increased, and the supplier's solution was to buy three new machines. To save schedule time, the NC programs were written before the machines were installed, but the size of the fixtures had been overlooked when specifying the machines.

The result was that 60% of the parts would not fit the new machines. To compound the problem, the older machines already owned, had specialized controllers that would not allow re-posting of the new NC programs; so the parts had to be programmed a second time to move them to older machines. All this gave the supplier a bad name and hurt schedules. Some of the parts were yanked from this supplier and placed in another machine shop—business was lost.

What is Required Part Accuracy (Drawing Specifications)?

All milling suppliers have their list of steps they take when considering a new or used machine acquisition, but I will list them anyway for completeness. For the parts you anticipate milling, determine the accuracy required and the part envelope sizes.

Study the drawing and its specifications to determine:

Machines

- Tolerances on part thickness at various cross sections, and part length.
- Hole patterns and true position requirements.
- Part curvature and flatness requirements.
- Part machined surface finish.
- Part envelope sizes (including fixture sizes.)
- Part material for anticipated dimensional changes due to variations in temperature.

Hole Patterns—True Position

Design engineers are constantly demanding greater accuracy from our machines. We recently had to find new milling sources for some eight-foot long by three-foot wide, curved (five-axis) parts. There are about 150 holes around the edge of the part with a true position requirement of 0.007 in. (.178 mm) when measured at the standard temperature of 68°F (20°C). Old machines and most new machines are not capable of holding this tolerance. Here we tested many machines to find the needed capability, but one cannot merely buy a machine based on the advertised machine accuracy (more on machine capability later).

Part Curvature Tolerances

The tolerances on curvature can usually be met on small parts, but the same tolerances can be very difficult on long parts. Doubling the length of a part more than doubles the tolerance problems.

Part Surface Finish

Surface finish of Ra 125 minch (0.0032 mm) or better is readily met with today's cutters and even older machines. Smooth surface finishes are generally achieved by special cutters and more importantly, by NC programming parameters.

Defining the Machine

I have separated out various topics to consider for discussion purposes. In reality, many of these categories are intertwined and overlap one another. For example, machine size, rigidity, and accuracy affect one another. Large flimsy machines do not yield the same accuracy as small rigid machines. The question is, "Will the machine under consideration yield the needed accuracy?"

Machine Type/Style

One of the most discussed parameters for a milling machine is the decision to buy a vertical or a horizontal machine. The arguments and trade-offs include chip evacuation.

Vertical vs. Horizontal

Proponents of horizontal machines like the fact that chips flush and fall away from the cutter readily. On vertical machines, especially in pocketing operations, one is often re-cutting chips that are not evacuated. The operator and a coolant fire hose can help this with special care, but this attention costs extra labor if the operator is trying to run more than one machine or is doing side operations on a previously milled part.

Proponents of vertical machines like the ease of setup and of part loading where you are not constantly fighting gravity. It often takes additional manpower to set up and load horizontal mills.

Some companies compromise by having a hinged machine bed on a horizontal mill. Here the machine bed is laid horizontally for setup and part loading, then elevated into a vertical position (usually with hydraulics) for the milling operations. This makes both camps happy. But it takes up more shop room for the bigger machine footprint, it is a larger capital investment, and maintenance is a bit higher due the greater complexity involved.

Spindle Speed

Many airframe milling shops buy machines that are intended to mill both titanium alloys and aluminum so it is important to discuss milling of both materials and the latest technical progress that is happening. The dichotomy between milling aluminum alloys and titanium alloys is growing.

Spindle speed is getting higher each month on milling of aluminum. We have some quasi-laboratory machines that run at 60,000 RPM and use 99-horsepower with 0.500 in. (12.7 mm) diameter cutters. Linear motors are used for the way drives, and we remove 500 in.^{3} (8.2 dm^3) of aluminum per minute. By the way, re-cutting of chips is not a problem at these speeds—chips are gone, but chip containment is a must. We use cameras to observe the cutting action because no one wants to volunteer to be anywhere near the cutter. The only reason we do not use this speed in regular production is poor spindle life. Current engineering of spindles limits us to about 20,000 RPM for reasonable spindle life in production, say one year or more of spindle life.

When milling titanium alloys we only needed about 2000 RPM until we developed high speed milling with cutters under an inch in diameter, but today we need nearly 4000 RPM to take advantage of the ability to finish mill at speeds up to 800 sfm (244 meters per minute).

The days when one could buy a universal machine for all metals is about gone. Competitors with specialized machines are more efficient. One option is to have a machine with several geared spindles that can operate at a wide range of RPM. Some of the more expensive machines can change spindles in a matter of seconds—much like changing cutters—and the NC program can do the spindle change. With these fancier spindles, one usually sacrifices rigidity and associated accuracy. The value of scrap chips is going up almost monthly. Therefore, chip management as one switches materials must also be considered in your overall decisions.

Nutating five-axis heads sacrifice some rigidity over three-axis heads, and some machines can automatically change from three-axis to five-axis heads by program to improve machine rigidity during three-axis milling.

Machine Accuracy/Capability

There are two main bodies of thinking on how to best determine a machines capability for producing parts: determinists and statisticians. Determinists base machine capability to produce accurate parts on the dimensional accuracy of the machine. Statisticians determine a machine or process accuracy by measuring parts produced under controlled conditions.

For determinists, determining machine accuracy is the most critical issue. Four major methods for measuring machine accuracy are noted by the researcher from on-the-

Chapter 5

job experience and are supported by Tobien (Tobien, T. "Development of measuring methods of large part verification at Cincinnati Machine." Cincinnati, Ohio. Author, 1999, pp. 15-39). The chronology is:

1. **Mechanical measurements of machines with precision rulers, micrometers, etc.** These means for determining machine accuracy have been used since 1850 when the first universal milling machines were built by Fredrick Webster Howe and Richard S. Lawrence in Windsor, Vermont. (Hubbard, Guy. 1923. <u>American Machinist, August 30, 1923, Volume 59</u>. Retrieved December 5, 2001 from the World Wide Web: (http://www.valley.net/~connriver/V09-60.htm). Mechanical progress on accuracy has improved consistently up to the present time.

2. **Machine mapping and certification by laser positioning.** Lasers and targets are strapped onto the machine bed for measurement of travel in the X, Y, and Z directions. Data is recorded and these records characterize a machine. Machines are re-certified in this manner once or twice per year. This method of machine measurement has been used for the last 30 years or so

3. **Laser tracking of spindle position.** A laser tracker is set up away from the machine spindle, a target is placed on the spindle, and spindle locations in the X, Y, and Z axes are mapped throughout the volume of the machine's working zone. The researcher's company has used this capability for the past eight years on very large drilling machines.

4. **Measurements with a Ballbar to provide a diagnosis of machine accuracy and repeatability.** These data provide an alternate method for mapping a machine. These devices have advanced to where they can provide electronic compensation files directly to the machine controller. This eliminates manual entry of compensation information.

Recent industry standards come into play, and are strongly used by determinists. Jim Destefani, Senior Editor of Manufacturing Engineering, 7/2001, cites Barry Rogers of

Machines

Renishaw Inc (Hoffman Estates, IL) as stating that:

> The purpose of ANSI/ASME B5.54 and ISO 230 [American National Standards Institute, American Society of Mechanical Engineers, an International Organization for Standardization] is not to specify the accuracy a machine must meet, but to find what accuracy level it can meet, that is, its process capability. Part prints dictate the accuracy your machine must have in order to make good parts. In short, the part print tells you where the accuracy bar is set: testing tells you how high your machine can jump.

The implication in this statement is that careful determination of machine accuracy can yield machine process capability. Producers of large machine tools are typically determinists. A recent report by Tobien of Cincinnati Machine Inc. cites a rule stating that machine tools and measuring instruments have to be at least four times more accurate than the parts being produced or measured (Tobien, T. "Development of measuring methods of large part verification at Cincinnati Machine. Cincinnati," Ohio. Author, 1999, p. 87). Tobien goes on to discuss the machine tool tolerances currently being held on major components for large machines:

- Parallelism = 0.032 mm (0.0013in)
- Flatness = 0.032 mm
- Perpendicularity = 0.025 mm (0.001 in)

These current machine tool tolerances[1] for large machines are not accurate enough to meet the researcher's company's current needs on special hole patterns with high accuracy requirements. Temperature variations are a major consideration on large parts because the machine tool itself expands and contracts with changes in temperature. When precision needs exist, large machine tools need to be characterized for their coefficient of thermal expansion.

---

1 If the 4:1 rule holds, these large machines will fail to meet the specification of 0.007-inch TP by a factor of 2:1 for the parts that inspired this investigation.

Chapter 5

Ratio Discussion (Machine Accuracy vs. Part Accuracy)

From previous work, I have never found a case where the accuracy of produced parts equals a machine's measured accuracy—a ratio of 1:1. On the contrary, I have found many cases where machine accuracy is not a direct indicator of process capability of parts produced. In a recent test, a machine that was advertised to be within 0.0002 inch (0.005 mm) of repeatable accuracy in all axes (X, Y, and Z) actually was only capable of holding 0.0020 inch (0.051 mm) of accuracy in any given axis during production. This finding was determined statistically from data on 24 parts, each containing five holes for a total of 120 holes. So, accuracy of this machine, relative to part accuracy in this specific case varied by a factor of 1:10. Three other large machines were recently tested in a similar manner. The parts produced exceeded the machine accuracy measurements by factors greater than 5:1.

During the researcher's early machine and tool design experiences at GMC Truck & Coach Division of General Motors Corporation, a one-tenth rule was applied (SPEAR Administration Staff. General Motors statistical process control manual, Warren, Michigan: General Motors: Author, 1986.). That is, the machine can only use one tenth of the part tolerance (1:10). Cincinnati Machine uses a one-fourth rule (1:4) [1]. One of our competitors uses a one-half rule (1:2); the production tool may use no more than one half of the drawing tolerance. At the researcher's present company, the rule is one third (1:3), and a March 2001 report from Independent Quality Labs, Inc. of Rockville, RI, reports that a one-fourth rule (1:4) is used today by many industries, especially for CMM measuring equipment. Roger Cope of Lamb Technicon indicates a need to maintain a ratio of 1:10 to assure part accuracy relative to machine accuracy (Cope, R. Manufacturing with 2020 intelligence. Manufacturing Engineering, 2001.). Lamb Technicon's main customer is the auto industry. The auto industry generally uses a one-tenth rule (1:10).

The lack of agreement between companies on what ratio rule to use weakens the stand taken by determinists. There may be some direct correlation between carefully determined machine accuracy and the accuracy of parts produced by the machine, but I did not find any confirmation of such a correlation. Reviews of current literature did not reveal agreement on which ratio to use. In fact, the reviews reveal variations in rules of thumb of from 1:1 to 1:10 as to what is the proper ratio of machine accuracy to part accuracy.

Machines

Further Comments. Certainly one cannot make an accurate part from an inaccurate machine. The machine must possess accuracy equal to or better than that needed for the parts being produced, but the antithesis that an accurate machine can produce parts that equal the machine's measured accuracy (1:1) is not necessarily true. Sometimes process capability is specified in purchase orders for new machines. Acceptance of the machine is then based on a demonstration of production of actual parts at either the equipment manufacturer before shipment or at the purchaser's facility. The point is that industry and I do not always accept the determinists' approach.

The most recent methods of measuring machine accuracy result in a determination of volumetric accuracy over the entire milling volume in some cases. This knowledge allows electronic compensation of several machine zones to be made in the machines controlling computer to improve accuracy.

Temperature Considerations

Is your shop air-conditioned? Two companies that I know without air conditioning have adapted their machines and processes with the capability of correcting the NC programs for temperature changes as they occur. NC programs are post-processed on the fly to compensate each tool path command for temperature. Temperature is measured in one shop by thermocouples attached to the part and fixture; in the other shop, the distance between tooling balls is measured. On-machine probes and software find and measure the distance between tooling balls every two hours. The fixture changes length with changes in temperature and amounts to a thermometer. Fresh programs are available for each range of temperatures encountered.

Therefore, the problem that CTE for part differs from CTE of the machine bed compounds the issue of process accuracy (Tobien, T. "Development of measuring methods of large part verification at Cincinnati Machine." Cincinnati, Ohio. Author, 1999, p. 83).

Part Size

Part size is not a major concern on airframe parts that are small, say less than 2 ft. (0.6 m) long. Modest temperature changes on small parts have very little effect on the part dimensions. On longer parts, like titanium alloy spars that are 8 ft. (2.4 m) long, a temperature swing of 6°F (3.3°C) can change the part length by 0.002 in. (0.051 mm)—or 20% of the tolerance if the tolerance is +/- 0.010 in. (0.254 mm). Aluminum, of course, grows in length about double the amount of titanium alloys for the same temperature change.

Part and Fixture Materials

It is important to select fixture material that has a coefficient of thermal expansion (CTE) that is about equal to the CTE of the part being held. I have seen parts warped beyond repair by a strong fixture made of non compatible material or, in some cases, fixture welds or fasteners broken from the stresses caused by temperature changes. The 30 ft. (9.1 m) diameter lifting fixture for the first stage of the Saturn Moon Rocket had six weld failures within minutes of moving it out of the factory into the hot Louisiana sun in 1965. The breaking welds sounded like rifle fire at close range.

My first choice of fixture material for titanium alloy parts is mild steel, as mentioned earlier in the chapter on fixturing.

Spindle growth in Z during warm-up

Some recent tests have shown that spindles grow in the Z or length direction of the spindle during warm-up on many machines. The length in Z can grow by as much as 0.002 in. (0.051 mm). This change in length is usually stable after two hours, so it might be necessary to pre-start the spindles a couple of hours before the shift begins to allow them to stabilize. We also had some cases a few years back where some high-speed ratcheting devices missed some strokes until they warmed up. The word was that they had morning sickness. When high precision is needed, watch for dimensional changes during machine warm up. Finding these dimensional culprits can be a difficult witch-hunt. One of our suppliers, striving for extreme accuracy, leaves the spindles running 24 hours per day to keep dimensions stable.

Coolant Temperature

Air conditioning by itself may not be enough to control part temperature. I have seen coolant change the part temperature by 8°F (4.4°C) in less than five minutes. The fire-hose action of heavy coolant flow won the temperature battle in this instance. See further discussion of coolant temperature control in *Chapter 8: Maintaining Large Part Accuracy*. The key here is to decide if you need the option of coolant temperature control on your machine. One company has a huge underground coolant tank that does not change temperature enough to worry about, but they did make many temperature measurements before relaxing on this issue. You need to determine your temperature needs and cost trade-offs. The point is that air conditioning is not automatically the most efficient answer to controlling part dimensions.

Special Accuracy

Laser Positioning and Thermocouples

For high precision airframe parts, one can install laser positional checkers on the X, Y, and Z axes. This works fine on a three-, four-, or five-axis machine, but the A and B axes of a five-axis machine cannot yet be controlled with laser positioners. Here the temperature of part and fixture are often measured with thermocouples on a continuous basis: the NC programming code is altered on the fly by a computer that feeds the machine's controller. I have seen the X dimension between holes that are about 18 ft. (5.5 m) apart held to within 0.002 in. (0.051 mm), when measured at 68°F (20°C), and this was in Southern California with the doors open and no air conditioning. Temperature compensation using coolant amounts to virtual air conditioning.

"The goal is to achieve enough machine tool accuracy to eliminate the need for drill jigs," (Zelinski, P., Shop with a nervous system. <u>Modern Machine Shop</u>, 1997.).

Use of laser positioning can allow a shop to operate with doors open and no air-conditioning. This laser positioning option initially cost up to $500 K per machine, but today costs about $80 K per machine installed. One machinery builder now offers laser positioning as a factory option. As an aside, I remember when turn signals and electric windshield wipers were offered as a factory options on cars. See Chapter 8 for further discussions on holding large part accuracy.

Laser Tracking

Laser tracking is another option for high accuracy, holes can be held within about 0.002 in. (0.051 mm) in both X and Y-axes. For this precision hole location on large parts

Chapter 5

(about 18 ft) (5.5m) long, we equipped several five-axis machines with a quill and with laser tracking and special software. A tracking ball was installed on the spindle housing about 12 in. (305 mm) from the cutter and a laser tracker was mounted about 15 ft. (4.6 m) away. When drilling via NC program, the spindle moves into position for drilling and stops for a second while the laser tracker activates and answers the question "is this the correct position for this hole." If the hole is not just right the machine nudges the spindle into an improved position—this action continues until the laser tracker says well enough—then the quill is activated to drill the hole. All this action is by NC program.

Glass Scales and How Mounted

Glass scales are another option for high precision. It was interesting to find out that when scales are longer than about one meter, steel scales are used, but everyone refers to them as glass scales regardless of length or material. Feedbacks from the read position on the glass scales cause slight adjustments in each axis for great positioning.

If the scales are anchored to the steel machine bed at the scale ends, there is distortion in overall length as temperature changes. One clever shop manager mounts these scales on a rather inert material that is not affected by temperature changes. The inert material is then mounted to the machine bed and is anchored only at the center of the axis being monitored. This gives a rather exact reading, much like the laser positioning described above because the machine CTE does not affect the length of the scale.

I never cease to be amazed at the ingenuity of the manufacturing community. There are a lot of clever people out there.

How I Handled a Special Accuracy Need

Sometimes special accuracy is required on part areas where interface with other parts is required or for special hole patterns.

Choice of the Variable to be Measured. The goal of one recent project was to find machines that can hold a tight tolerance (0.007 in. true position on hole locations in large contoured parts. The required tolerance here was 0.007 in. (0.178 mm) true position for the holes.

Literature shows that the statistical control chart methods using variables for predicting process capability are more accurate than the methods using attributes (*p* and *Np* charts.) "Variables data have the accuracy advantage because they reveal how close we are to a specification —." (Cooper, D. R., & Schindler, P. S. <u>Business Research Methods: Sixth Edition.</u> Irwin McGraw-Hill, 1998, p. 631).

The researcher could not find a single instance of control charts being employed for predicting the process capability for fabrication of large parts.

Machine Wear Considerations

Nearly all milling machines recently marketed will hold at least +/- 0.003 in. (0.076 mm) of accuracy on large parts even after several years of machine wear. This accuracy implies proper maintenance and annual or semi-annual re-calibration of the machine. This accuracy is usually good enough because dimensions on most large airframe parts are specified by drawing to be held to +/- 0.010 in. (0.254 mm).

Another importance of knowing a machine's accuracy is to allow monitoring of machine degeneration due to wear or damage. The literature reviewed mentioned this importance, but did not show the correlation of machine wear to accuracy of part produced. This correlation to machine wear is not pertinent to this book, but may be a reason for future research.

Machine Features and Functional Characteristics

Vertical vs. horizontal mills were discussed earlier.

Number of Spindles

Some shops only own single spindle machines. Other shops have machines with anywhere from three to eight spindles. I work with large airframe parts where quantities are low. Perhaps we build 200 airplanes of a given model in ten years. How should one equip such a shop? Certainly every shop has its own culture to work with when it comes to calculating how many spindles per machine or per gantry are most efficient for you. My suggestion is that you do not simply copy your neighbor shops in this regard.

To help get my thinking straight I like to remember the days when I carpooled to work. If two people car pool your costs are cut by about 50%. If three persons carpool, the initial cost is reduced by about 67%, but the gain from going from two to three persons is only 17%. Similarly, the additional savings from going from three to four persons is a paltry 8%. I say paltry because along with a small gain in savings after two people comes the extra aggravation of having everyone prompt and ready at the appointed time. The risks of delays for that extra person get higher fast with the number of persons trying to pool together. In fact, if one puts a dollar value on the aggravation and complexities of having several people in the car pool, there may not be any savings at all when you get above two or three people.

I use the carpool example because the same kinds of elements are present when you calculate the returns for machines with various numbers of spindles. For example, when you run parts on a multi-spindle machine what percentage of the time do you have enough raw material to utilize each spindle? Many times when we run parts, it is a bit like buying six hotdogs in a package and eight buns in another package. It doesn't always match up. If you use four spindles, you need to build four fixtures and then spend extra time aligning them every time you set up. Most of my sample calculations show that the correct answer to the question of "how many spindles do I buy?" is usually two or at most, three spindles. When you go beyond two spindles you begin to give up a lot of flexibility toward future improvements and drawing changes, and the savings are not that much greater.

Number of Axes

Once again I am thinking about milling titanium alloys. Three-axis machines are more rigid and are cheaper than five-axis machines; therefore, many parts are rough machined on three-axis mills even though final milling requires five-axis equipment. Please also read my discussions in Chapter 7 on high-speed milling of titanium alloys where I discuss the need for five-axis roughing in the final media before finish milling, if you want to finish mill faster than 400 sfm (122 m/min).

Fourth and Fifth Axis Flexibility (Angles)

Most airframe spars and frames today are five-axis designs. If the greatest part angles are less than 25 degrees, then the older five-axis gantry machines can do the job.

Machines

Today's five-axis multi-spindle gantry machines can usually move 30 degrees in the "A" and "B" axes, and some are even 35-degree machines.

If more angular flexibility is required without special rocking or multi-positional fixtures or use of angle plates, then you need to consider single spindle machines with nutating spindle heads. These machine spindles can rotate to angles beyond 90 degrees, but once again some rigidity is lost.

Gantry Risers and Tables

Risers under gantries and riser tables allow access to both short and tall parts on the same machine. Until recently, airframe parts were only one-dimensional or two-dimensional parts. I define a one-dimensional part as a long slender part such as a wing spar where the long or X dimension of the part is the only important dimension that you need to know when selecting a machine to mill the part. A two-dimensional part has large dimensions in both X and Y like an aircraft frame member, but still no major height in the Z direction. These one and two-dimensional parts were then fastened together to create the framing for wings and fuselages.

Because there were no three-dimensional parts (parts with a high Z dimension), airframe machine shops did not own production machines that could take a part with a Z dimension greater than about 18 in. (457mm). We had over 100 three-spindle gantry mills with a typical Z dimension of 26 in. (660 mm) from the nose of the spindle to the bed of the machine. Today we are welding sections of an aircraft fuselage together rather than using fasteners and we now require machines with 5 ft. (1.5 m) or more in the Z axis to handle these three-dimensional parts. An inexpensive way of modifying a gantry machine with only 26 in. (660 m) in Z is to put risers under the gantry legs. This allows three-dimensional parts that require greater than 26 in. (660 m) in Z to be fixtured and machined with existing machines (provided the total travel in the Z axis is small.) Many new machines are purchased with the risers already in place.

You can also purchase riser tables for use when milling parts on a machine with risers, but it may be cheaper to make your own tables. About half the companies that use riser tables make their own tables.

Chapter 5

What about milling one- and two-dimensional parts on this now tall machine? The answer here is to use riser tables that in essence create an artificial machine bed closer to the spindles. This is a nice solution that allows old programs and fixtures to be utilized without sacrificing rigidity. If a spindle plus cutter extends too far in Z, then rigidity is lost—the riser tables solve this dilemma, and are ergonomically better for the operator. It's kind of like working with raised flowerbeds—not much bending and kneeling is required.

Controllers

It is hard to predict the future, but you do need to anticipate the future when selecting controllers. It is also nice to have common controllers on all machines for the sake of operators and for NC programmers.

Two examples of what did not work well. A major supplier of milling purchased a machine shop about five years ago. Half of the machines had special home-built controllers from about 20 years ago, and half had industry standard controllers. No one has figured out how to re-post jobs when they want to move to the other type of controllers. This means one needs to rewrite the NC programs, re-check them, and sometimes even produce a part for special first-part inspection to prove out the new process. There is no flexibility; whatever savings were anticipated from using the specialized controllers is more than offset by the cost of re-doing everything. In another shop, we purchased a family of controllers and installed them on all our machines. But within a couple of years, the controller manufacturer decided to quit building and updating these controllers. We thought we had properly predicted the future, but it did not work out.

Chip Removal

Two major areas need to be considered in chip removal:
1. How to remove the chips from the machine and cutting area—we especially do not want to re-cut chips and thereby reduce cutter life. On titanium alloys, the reduction in cutter life when re-cutting chips is about 35%. This figure is based on some studies I did on cutter life about 10 years ago.
2. The other major consideration is ease of keeping chips from different metals separated. This issue is even more important today because of the high price for

properly-handled chips.

Sale of chips that are properly handled can be considered as another product sale at today's prices.

Coolant Considerations

Low coolant pressure (about 100 pounds per in.^{2} {690 kPa}) is the norm today on about 98% of the machines that machine titanium alloys for our aircraft, and through-the-spindle coolant is used on about 30% of the mills. It appears that the use of low-pressure coolant will be with us for a considerable time, and through-the-spindle use is growing.

Through-the-spindle coolant is especially important when performing rough cuts because here the cutter is often buried in the work piece and through-the-spindle coolant helps flush the chips away. Through-the-spindle coolant is not as important for finish cuts where the cutter is only making partial contact for much of the cutting; here, the cutter teeth are cooled externally by the flood of coolant.

If through-the-spindle coolant is not available on a given machine, one can achieve it by using a clamshell device around the tool holder to get through-the-spindle coolant. Operators hate the extra work of using a clamshell, but sometimes it saves money and time. The extra time spent can be a proper investment.

Maintenance and Preventive Maintenance

Maintenance and preventive maintenance covers: areas of preventative maintenance, frequency of what kind of checks (dimensional and otherwise), what repair parts to stock, grinding of spindles in-place, and what spindle run-out is allowed?

I recently saw one shop borrow a spindle motor from a competitor until the replacement became available three weeks later. I say working out ways of sharing with a competitor is not only possible, but it is a good thing to do (see *Chapter 16: Inter-Company Sharing*).

Chapter 5

Process Considerations

Part Setup

Loading and unloading of parts is discussed earlier in Chapter 4.

Machine Bed Flush with Floor

One shop that I visit on a regular basis installed new large gantry machines so the machine bed is about parallel with the surrounding shop floor. This allows loading of fixtures and parts with a forklift truck, which can drive out onto the bed via a metal plate (bridge). This can be very helpful in a new plant where 1) the cranes may not be installed until a year or two later, 2) portable crane trucks need to be rented, and 3) we get delays when they are not immediately available. Another advantage is, if your cranes break down. It does not add much to the cost of the special foundation to move it down about 18 in. (457 mm).

Future Machines

My Predictions on Future Machines

We will see horizontal mills that machine two sides of a part at the same time. Here, the part will be fixtured vertically between the heads. In thin part sections, one head could become a backing pad, opposite the cutting head, to allow a very thin wall to be milled—thin meaning thinner than 0.040 in. (1.016 mm) because we can achieve 0.040 in. thickness without support tooling. With such a moving support pad we would have what amounts to a live and programmable fixture. I believe this is a fresh concept.

How about inventing a machine with a vision system located near the spindle that can measure the distance to the part surface and allow program revisions on the fly? It would be great to know the exact terrain of the part surface when trying to countersink for perfect fastener flushness or when trying to mill a seal groove whose importance is depth of groove relative to the part surface on some five-axis part.

If we could optically measure surface finish on the fly, why not speed up the feed until the achieved finish becomes close to the drawing specified finish of, say, Ra125 minch (0.0032 mm)? If the feed rate would vary to give a desired finish, it would save perhaps 30% of the milling time because today we generally give the customer a better finish than is being paid for.

How about after a machine crash, have built-in software to assist in recalibrating a machine in minutes instead of hours and days?

Your Role in the Future of Machines

By feeding parameters to the machinery builders like those mentioned above under future machines, we can begin to influence the machines of tomorrow. How much stronger would our requests be if they were echoed by other shops with whom we have learned to share technology?? What's wrong with wanting to change the world just a bit?

GROUP III: MILLING OF TITANIUM ALLOYS

Rough Milling of Titanium alloys, Finish Milling of Titanium alloys, Maintaining Large Part Accuracy, Milling Difficult Features, and Warpage.

Here we get to the core of this book—making chips in the most efficient manner possible. I see lean studies as primarily focused on minimizing the non-value-added portions of a process. These chapters work on the value-added portions of the process. The first two chapters in this group are presented in the normal sequence of milling operations: *Chapter 6: Rough Milling of Titanium Alloys*, then *Chapter 7: Finish Milling of Titanium Alloys*.

Problem milling areas that seemed important enough to warrant their own chapters are then presented in Chapters 8–10. *Chapter 8: Maintaining Large Part Accuracy* discusses the problems of maintaining large part accuracy in the face of temperature changes, and includes a section on my prediction that drill fixtures are essentially obsolete for large parts. *Chapter 9: Milling Difficult Features* covers milling of thin webs without the need for support tooling and gives parameters of how to machine a specific rough surface finish. *Chapter 10 – Warpage* covers the problems of warpage and distortion along with some thoughts on straightening parts, but most importantly this chapter discusses measures you can take to prevent warpage and distortion.

Definitions

Carbide cutters—refers to solid carbide cutters or carbide inserted cutters made of micro-grain or sub-micro-grain carbide which means the grains measure no greater than 0.5 microns.

Climb milling—refers to the direction of cutter rotation where the teeth enter into the top surface of the material rather than entering under the chip and lifting, as in conventional milling which is the opposite rotation.

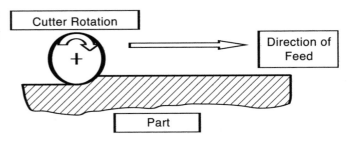

Cobalt cutters—cutters made of a good grade of high- speed steel with 8–10% of cobalt; usually M-42.

Conventional milling—in conventional milling, the cutter rotates in the opposite direction of climb milling. Here the cutter teeth enter under the chip and lift the chip. Conventional milling was the norm until about 40 years ago, but unfortunately is seldom used today

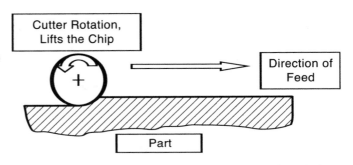

High-speed milling of titanium alloys—moving beyond the conventional speeds of 60 sfm (18 m/min.) for cobalt cutters and 120 sfm (36.6 m/min.) for carbide cutters.

HRC (Hardness Rockwell C scale)—needs some explanation when applied to titanium alloys. Titanium hardness is usually not specified because readings vary by several points depending on where the reading happens to fall—so when I refer to the **HRC** of titanium alloys, I am referring to an average of at least ten readings.

Spiral or trochoidal milling—a cutter path that is similar to the path made by a mark on the side of a wheel as it rolls along a surface. The center of the cutter spirals along such a path.

TiAlN—cutter coating—Titanium Aluminum Nitride.

Titanium alloy, ti-6Al-2Sn-2Zr-2Cr-2Mo-STA—the STA stands for Solution Treated and Aged. HRC 43 average.

Titanium alloy, ti-6Al-4V-ELI-BA—the ELI in the alloy description stands for Extra Low Interstitial and in simple terms this means that less oxygen is allowed in this alloy, which gains material toughness but loses some strength in the trade-off. The BA at the end of the alloy description stands for Beta-Annealed (heat treated) HRC 41 average.

Chapter 6

ROUGH MILLING OF TITANIUM ALLOYS

The purpose of this chapter is to present the latest techniques for high-efficiency rough milling of titanium alloys. Little progress had been made in rough maching of titanium alloys until February of 2004.

This book does not get into the details of standard and sound milling practices for general machining such as the notion of keeping at least two cutter teeth in the work. These practices are well described in many other publications, and it is presumed that your machinists are well versed in these general milling methods.

This chapter opens with a discussion of the traditional practices for rough milling of titanium alloys. Limitations caused by chatter are discussed. This is followed with some recent breakthroughs in rough milling that reduce milling time by about one-third.

Traditions and Basics for Rough Milling of Titanium Alloys

For the past 40 years, nearly all rough milling of titanium alloys has been with cobalt cutters (made from M42 cutter material) at speeds up to 60 sfm (18 m/min..) With 0.005 in. (0.127 mm) chip loads, this gives a minimum of 1 hour of cutter life even with the cutter buried in the material. Chip loads per tooth that vary from 0.002–0.012 in. (0.051–0.305 mm) are used. If chip loads of less than 0.002 in. are used, there is danger of work hardening and of overheating.

The traditional cutter material for roughing of titanium alloys is (M42). Cobalt cutters are used because carbide does not hold up well on the uneven surfaces of castings and forgings. The exception is to use cutters such as carbide insert face mills on titanium alloy plate where the surface is known to be even and flat.

Some tests performed by me about ten years ago showed about 35% greater cutter life with TiN (Titanium Nitride) and Tin2 **coated** cobalt cutters, but the cutters cost more if coated and you need to re-coat after re-sharpening—so there is little payback for coating of cobalt cutters used for roughing based on the TiN coatings that I tested.

Standard Parameters for Rough Milling

The following parameters for rough milling with uncoated cobalt (M42) cutters will give low cutter distortion and minimize chatter:

Use 40% of cutter diameter for radial engagement

Use 100% of cutter diameter as the figure for axial depth-of-cut. This ratio of cutter diameter to axial depth-of-cut is to minimize cutter distortion or bending during the cut.

There have been many tests for the following size cobalt (M 42) positive rake cutters and chip loads:
- 0.250 in. (6.35 mm) diameter—4 flutes—0.004 in. (0.102 mm) chip load/tooth
- 0.375 in. (9.525 mm) diameter—4 flutes—0.005 in. (0.127 mm) chip load/tooth
- 0.50 in. (12.7 mm) diameter—4 flutes—0.005 in. (0.127 mm) chip load/tooth
- 0.750 in. (19.05 mm) diameter—4 flutes—0.006 in. (0.152 mm) chip load/tooth
- 1.000 in. (25.4 mm) diameter—6 flutes—0.006 in. (0.152 mm) chip load/tooth
- 2.000 in. (50.8 mm) diameter—8 flutes—0.008 in. (0.203 mm) chip load/tooth

Basic Values for Cutter Life vs. Sfm

Using the above parameters, testing shows that cutter life for roughing titanium alloys with solid cobalt (M 42) cutters yields a very simple set of variables. Cutter life vs. sfm values can be plotted as a curve depicted by these values:
- 72 sfm (21.9 m/min) yields 30 minutes of cutter life.
- 60 sfm (18.3 m/min) yields 60 minutes of cutter life.
- 45 sfm (13.7 m/min) yields 90 minutes of cutter life.

Rough Milling of Titanium Alloys

- 38 sfm (11.6 m/min) yields 5 hours of cutter life.
- 30 sfm (9.1 m/min) yields 8 hours of cutter life.

These approximate values for cutter life hold for cobalt cutters in the diameter ranges listed above. The curve for the first three sets of cutter life data is shown for both US and metric values as shown in Figure 6-1 and Figure 6-2, respectively.

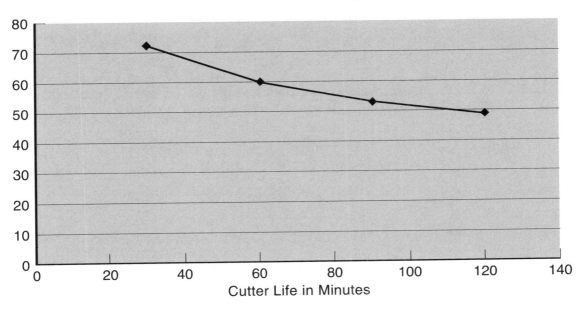

Figure 6-1: Cutter Life Curve For Rough Milling

The secondary variable that affects cutter life is chip load. I do not have quantitative values for variations in chip load, but have seen enough data to know that it affects cutter life.

When cutting pockets in titanium alloys with a vertical mill, we have the problem of recutting chips. With the pocket filled with coolant, it is often difficult to see the chips and to see what is happening. I did some testing of cutter life under these conditions of recutting chips a few years back. My findings were that cutter life was reduced about 35%

Chapter 6

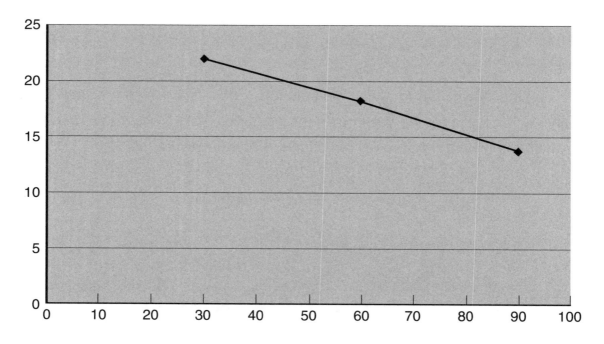

Figure 6-2: Cutter Life Curve For Rough Milling (Metric Values)

when chips were being recut in these pockets. The solution is to make special efforts to flush away the chips. Normal flood coolant is not enough to solve this problem. We need to use "fire hose" techniques which cost extra operator effort. In one shop a pair of 0.500 in. (12.7 mm) diameter tubing is used next to the vertical spindle with a heavy flow of coolant. This acts as a rather permanent "fire hose" and frees the operator from this task. Two of these coolant lines are used at each spindle, and the coolant is supplied with a separate pump.

Use of cobalt **wave cutters** or similar functioning cutters for roughing also produces good results when cutting pockets with a vertical mill. The reason wave cutters work well is that they produce a light weight chip that generally floats away with strong coolant flow. This eliminates most recutting of chips when vertical milling. Strong coolant flow is flood coolant with about 100 pounds/in^2 (690 kPa).

Rough Milling of Titanium Alloys

Chatter—A Primary Limiter to Metal Removal When Roughing

When chatter is encountered while milling titanium alloys, the speed of removing chips is usually limited in order to maintain proper surface finish and good cutter life. The main cause of chatter when milling titanium alloys is vibration within the part, fixture, and tombstone. While detecting and solving harmonic vibrations in the spindle/tool holder/cutter combination works well to solve chatter problems when cutting aluminum at high RPM, the same solutions do not usually help with milling titanium alloys. The solution to chatter when milling titanium alloys is to use rigid milling machines, rigid fixtures, and to dampen and rigidize the fixture/part/tombstone. The use of cutters with uneven division of cutting teeth also helps to minimize chatter in specific cases.

Efficient Rough Milling Today

Most rough milling of titanium alloys is accomplished using the traditional methods described above. These are the tried and true methods, but on selected parts there are sections of material that can be more efficiently removed with the methods described below.

Three areas of improvement in rough milling of titanium alloys have developed—all within the past two years. The primary alloy involved thus far is ti-6Al-4V-ELI-BA. The three areas of improvement are:
- Plunge roughing with carbide insert cutters.
- Trochoidal (spiral) side milling with solid carbide cutters. Solid carbide cutters are made of micro-grain or sub-micro-grain carbide.
- Use of powdered metal cutters.
- Application of these techniques for rough milling of titanium alloys have resulted in milling time reductions of about one-third on selected parts.

<u>Plunge Roughing</u>

Plunge roughing brings carbide inserts into the rough milling picture for titanium alloys. The use of plunge roughing is new for titanium, but is not new in the industry having been used for years on other hard metals such as die milling.

Chapter 6

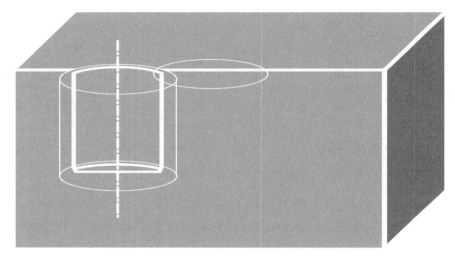

Figure 6-3: Plunge Roughing

Plunge roughing started being used in milling of titanium alloys in the past few years. It is done with plunge cutters that are as large as 4 in. (101.6 mm) diameter. Plunge cutters for plunging should have as many inserts as possible. Inserts are on their side for end cutting, and through-the-spindle coolant is desirable when vertical milling to get chips out of the way (see Figure 6-3). Flood coolant also works well. The plunge cut is usually at a feed rate in the Z axis of 20–40 in./min (508–1016 mm/min)—this reduces roughing time by nearly 67% over conventional roughing with cobalt cutters.

We use a rule of thumb of 0.5 gallons/min./horsepower (1.9 l/min./horsepower) of coolant flow through the spindle. We use a radial depth of cut of about 0.5 in. (12.7 mm) and 0.65–1.250 in. (16.5–31.75 mm) step sideways. The radial depth of cut varies with the insert face length. Use a modified drill cycle with a slight move out (radially) before retraction. The radial move away from the cut will reduce cutter wear by about 35% because the radial move out prevents the inserts from rubbing during cutter retraction. An option to the radial move out is to reposition the plunge radially for the next lower step on waterfall types of cuts. We use 400–600 sfm (122–183 m/min). With 6 inserts this yields about 18 in^3/min. (0.295 dm^3/min). Note: horsepower comes into play here—allow at least 1.4 HP per cubic inch of titanium removal per minute. I have seen machines stall on two occasions. A tip to minimize cutter breakage is to use a programming feed code that relates feed rate to feed/revolution. This might save your cutter if the machine begins to stall or slow. Note also that torque doubles when the cutter wear approaches 0.005 in. (0.127 mm) wide on the sharp cutting edge. The horsepower will of course double as well—30 horsepower now becomes 60 horsepower and can stall the machine.

One supplier uses a 4 in. (101.6 mm) diameter cutter, step 1 in. (25.4 mm) radially into work and 2.5 in. (63.5 mm) sideways, 6 inserts at .008 in. (0.203 mm) chip load per tooth and 180 sfm (55 m/min) with flood coolant. This yields about 9 in.^{3}/min. (0.147 dm^3/min).

Trochoidal (Spiral) Milling

Solid Carbide Cutters: Spiral milling (Trochoidal milling): Spiral milling is being used by one of our machining suppliers. Here solid **carbide** cutters are used for rough milling. We generally only used **cobalt** cutters for roughing until the last two years—see details in Spiral milling below.

We have begun doing some spiral side milling during roughing operations (also known as Trochoidal milling)—best done with cutters that are under 1 in. (25.4 mm) diam-

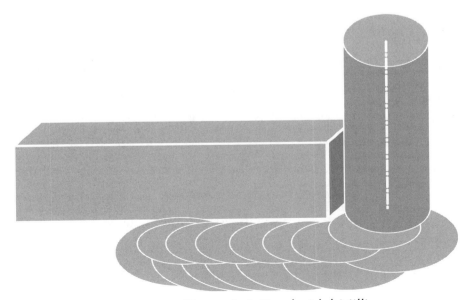

Figure 6-4: Trochoidal Milling

eter. These cutters are solid carbide coated with TiAlN. The sfm used is as high as 600, chip loads are up to 0.006 in. (0.152 mm), and radial depth of cut of 0.030 in. (0.762 mm) maximum. Limiting the radial depth of cut keeps most of the teeth out of the material and allows the coolant to keep these teeth cool—hence longer cutter life.

Spiral or trochoidal milling (see Figure 6-4) is best done with cutters that are under 1 in. (25.4 mm) diameter. Use solid carbide coated with TiAlN, a 0.250 in. (6.35 mm) spiral motion, 600 sfm (183 m/min), and climb milling. This calculates out to 3600 RPM for a 0.625 in. (15.875 mm) diameter cutter.) Chip load is 0.004–0.005 in. (0.102–0.127 mm) per tooth. Usually limit Axial Depth of Cut (ADC) to one diameter. Care is taken to have no more than 30 degrees of cutter engagement maximum. This allows two great things to happen:

- Allows cutting of rough—as forged—surfaces with solid carbide cutters instead of the slower cobalt (M42) cutters.
- Keeps the cutter cool
- A second use of spiral (trochoidal) cutting is in pocket corners. Here we come back with a smaller diameter cutter and spiral mill the corners to finish them. All these cuts are climb milling.

Powdered Metal Cutters

Don't jump into this one. I had some bad experiences with powdered metal cutters about ten years ago. Powdered metal cutters of that era did not hold up on parts that had hardness above HRC 40. One of our most commonly used titanium alloys is ti-6Al-4V-ELI-BA—it averages about HRC 41, and less commonly used is ti-6Al-2Sn-2Zr-2Cr-2Mo-STA which averages about HRC 43. Recent progress in powdered metal technology has caused me to re-look at powdered metal cutters for milling titanium alloys. Some recent applications of powdered metal cutters have reduced roughing costs. 70 sfm (21 m/min) is being used for powdered metal vs. 60 sfm (18.3 m/min) for cobalt with about 1.5 hours of cutter life. Two of our milling suppliers are having some success working with powdered metal cutters. The powdered metal materials for these cutters are ASP#60 from

Rough Milling of Titanium Alloys

Hanita (contains 12% cobalt) and Rex 15, 20, and 75 from other suppliers for rough or finish cutting. It is important that these powdered metal cutters are coated with TiAlN. Some other recent tests with powdered metal cutters are not working out so well (cutter life of less than 10 minutes at 70 sfm (21.3 m/min). One needs to try powdered metal cutters carefully for your specific milling needs in titanium alloys.

Summary

Traditional rough milling of titanium alloys is to use cobalt (M42) cutters with positive rake. These cutters are run at 50–60 sfm (15.2-18.3 m/min) with and chip loads per tooth that vary from 0.002–0.012 in. (0.051–0.305 mm). The primary variable that affects cutter life is sfm, and chip load per tooth is the secondary variable that affects cutter life.

The primary causes of chatter when milling titanium alloys are in the machine rigidity, fixture, part, and tombstone—so the solution to chatter problems when milling titanium alloys is to use rigid milling machines, rigid fixtures, and to dampen and rigidize the fixture/part/tombstone.

- The three areas of recent improvement in rough milling of titanium alloys are:
- Plunge roughing with carbide insert cutters.
- Trochoidal (spiral) side milling with solid carbide cutters.
- Use of powdered metal cutters.

Application of these techniques for rough milling of titanium can result in milling time reductions of about one-third on selected parts.

CHAPTER 7

FINISH MILLING OF TITANIUM ALLOYS

"The rule of thumb in tooling development, is make sure the cutting tool is strong enough first, then worry about speed," Hector Davis, Forrest Machining, Inc., one of our machining suppliers.

The purpose of this chapter is to review traditional milling parameters and to present the latest techniques for high speed milling of titanium alloys—techniques for increasing the speed of finish cuts by a factor of up to 8X are discussed. Cutting efficiency depends not only on spindle speed, but on the type of material, depth of cut, and cutter length and diameters.

Traditions and Basics for Finish Milling of Titanium Alloys

Traditional finish milling of titanium alloys has been accomplished with both cobalt (M42) and carbide (micro-grain or sub-micro-grain carbide) cutters and climb milling. The normal amount of material left for cleanup was usually 0.100 in. (2.54 mm).

The cobalt cutters (made from M42 cutter material) were run at speeds up to 60 sfm (18 m/min..) Chip loads were typically about 0.005 in. (0.127 mm), but varied from 0.002–0.012 in. (0.051–0.305 mm). These cutting parameters typically give a cutter life of about 90 minutes. Remember that if chip loads of less than 0.002 in. are used there is danger of work hardening and of overheating which will reduce cutter life.

Chapter 7

Traditional finish milling cuts on titanium alloys are also made with solid carbide cutters and with carbide insert cutters run at 100 to 120 sfm (30–37 m/min), and with typical chip loads of 0.005 in. (0.127mm). The decision of whether to use cobalt or carbide cutters is influenced by:

- Availability and cost of cutters
- The evenness of surfaces—carbide cutters do not last long in intermittent cuts.
- Carbide cutters remove metal about twice as fast as cobalt when using traditional cutting parameters.

Efficient Finish Milling Today

Breakthrough progress in milling titanium alloys at high speeds was made at Boeing about eight years ago, and almost simultaneously, Hanita Cutting Tools Company made similar findings in their laboratories. The key to these improvements is to leave less material for finish milling—a maximum of 0.030 in. (0.762 mm) for cleanup instead of the traditional 0.100 in. (2.54 mm). On finish cuts, this allows us to go from 120 sfm (36.6 m/min) using carbide cutters to 600 sfm (183 m/min). **For finish milling of titanium alloys, we now operate at feed rates as high as 40 in./min (1016 mm/min); the old speed was 2 in./min (51 mm/min). These feed rates equate to the 600 sfm (183 m/min) discussed above.**

In our progress toward higher speeds, it seems that the reason tool life is good (usually above one hour) is because the teeth of the cutter are only engaged a tiny amount of the time—most of the time the teeth are out of the metal being washed with coolant.

Leaving less material for clean up allowed us to mill up to 400 sfm (122 m/min). To achieve 600 sfm (183 m/min), we did two things:

We went to five-axis milling on the roughing or intermediate cuts that preceded finish milling. This five-axis intermediate cutting minimizes the intermittant cutter loads that ruin carbide tool life.

When cutting with the bottom of the cutter, we tilted the cutter about ? to ? degrees to lift the heel of the cutter that is doing the bottom cutting.

The five-axis roughing meant we were leaving a very consistent amount of metal for finish cuts. Solid carbide cutters do not like inconsistent surfaces.

Tilting the cutter gets the heel of the cutter up where coolant can do its cooling. Without tilt we generated too much heat—the bottom of the teeth were always in contact with the metal and we could not machine faster than 400 sfm (122 m/min) without sacrificing cutter life.

One more step is often required. On side cutting, we need to re-run the final milling pass because the cutter is not rigid enough to meet our nominal dimensions. In theory, one could build in compensation for cutter deflection and eliminate the extra pass, but this is a tough process to manage. If the operator pauses the machine for any reason during a compensated finish pass, the cutter will tend to walk into the part beyond nominal dimensions and we have scrap. On deep cuts, it is much safer and easier to manage high-speed milling if we use a spring or wash pass.

High-speed milling of titanium alloys is all about cutter life, and heat is the enemy. So once again I tout the use of sharp cutters to keep heat down. In high-speed milling, we sometimes leave only about 0.025 in. (0.635 mm) of material for the finish milling and use solid carbide cutters that are 1 in. (25.4 mm) diameter and smaller.

The discussion on cobalt cutters for roughing operations included findings that coatings usually gave us about 35% greater cutter life at the old traditional feeds and speeds, but the coated cutters cost about 30% more if they are coated, so it didn't pay to coat. Also we didn't have to re-coat after re-sharpening. Today, our testing of coatings—TiAlN (Titanium Aluminum Nitride) on the solid carbide cutters used for **finish** milling at high speeds—400–800 sfm (122–244 m/min) results in nearly double the cutter life when compared to uncoated carbide cutters. We also re-coat after each re-sharpening.

One of our suppliers utilizes finish milling cuts on titanium alloys that are now as high as 800 sfm (244 m/min), and their goals are even higher, but most machining suppliers are using about 400 sfm (122 m/min) to get longer than 90 minutes of cutter life.

The biggest road block to achieving these higher speeds on finish milling of titanium alloys has been convincing the NC programmers to program for these higher speeds. Another major inhibitor to achieving higher titanium alloy cutting speed was the reluctance of older machine tool operators to run at the newer levels.

Some further thoughts:
1. Take bottom cuts in pockets to net dimensions during roughing and avoid having to tilt the cutter during finishing because the pocket floor is already finished.
2. Program the finish-milling pass to be about 0.001 in. (0.0254 mm) above the pocket floors and eliminate the need to bottom cut.
3. Another possibility for saving time would be to write a separate program for the final spring pass and run it up in the 1000 to 1200 sfm (305–366 m/min) range. We are removing less than 0.002 in. (0.0508 mm) of material during this spring pass, the cutter should handle this speed, and so what if the cutter only lasts about 15 minutes at the higher speed.

Further Thoughts on finish Milling

The impossible is happening. One of our milling suppliers is getting high-speed finish milling success on titanium alloys with cobalt (M42) cutters. All our tests were with carbide for finish milling. We did some fresh testing of cobalt at high speeds on finish cuts where 0.025–0.030 in. (0.635–0.762 mm) was left for clean up. We had success with cobalt at about 400 sfm (122 m/min) with 0.005 in. (0.127 mm) to confirm the supplier results.

To me this result is ironic: the supplier heard about high-speed finish milling of titanium alloys, but did not know that one **"has to switch to solid carbide"** for high-speed results. We experts that developed high-speed cutting with solid carbide cutters **presumed** that cobalt would not work at high speeds when removing less material (0.030 in. [0.762 mm]) and never bothered to test cobalt at high speeds until now. TiAlN coatings were used. Sometimes we are not as smart as we think we are. Traditionally, cobalt cutters on **heavier** cuts will only give reasonable tool life of one hour or more when driven at 50–60 sfm (15-18 m/min).

Solid carbide and carbide inserts are the primary cutting materials for high efficiency finish cuts at 600–800 sfm (183–244 m/min), but we now find that cobalt cutters can be used for finish speeds as high as 400 sfm (122 m/min) vs. the traditional 60 sfm (18 m/min) provided the radial depth of cut does not exceed 0.030 in. (0.762 mm). This allows most of the teeth to be cooled while not in the cut.

Extending Cutter Life

Solid carbide cutters—Several milling suppliers are experimenting with new coatings such as "X.ceed" from Balzars and a new formulation from Hanita. Early reports show as high as 3 hours of tool life at 600 sfm (183 m/min) when side cutting. (Note: Tool life for carbide cutters is 45 minutes with no coating and 1.5 hours with TIALN coating at these speeds).

Summary

Traditional finish milling of titanium alloys has been accomplished with both cobalt (M42) and carbide (micro-grain or sub-micro-grain carbide) cutters and climb milling. The normal amount of material left for cleanup was usually 0.100 in. (2.54 mm).

Leaving less material for finish milling—a maximum of 0.030 in. (0.762 mm) for cleanup instead of the traditional 0.100 in. (2.54 mm) allows us to go from 120 sfm (36.6 m/min) using carbide cutters to 600 sfm (183 m/min). **For finish milling of titanium alloys, we now operate at feed rates as high as 40 in./min (1016 mm/min); the old speed was 2 in./min (51 mm/min).**

CHAPTER 8

MAINTAINING LARGE PART ACCURACY

The discussion on maintaining large part accuracy will be handled by folllowing the development of high accuracy for a specific family of parts.

This chapter begins with a discussion of temperature control and goes on to discuss the following areas of resolution of a specific accuracy problem:
- Statement of the problem
- Drill jig accuracy and limitations
- Development of precision drilling centers
- Finding other precision machines to handle production overload
- Development of a test fixture to predict machine capability
- Results and findings of testing
- The chapter concludes with a brief summary.

Definitions

Drill Jig: Tooling fixture with bushings for each hole location that guides drilling operations. Drilling is usually accomplished with portable drills.

Large parts: For purpose of this report, large parts are parts that measure a minimum of two meters long.

True Position (TP): See the following sample calculation of TP
- TP is the *diameter* of the circle in which the center of the produced hole must fall.
- *Diameter* equals *radius* times two.
- *Radius* equals the hypotenuse of the triangle that has two sides of 0.010 in. (0.254 mm).
- So radius = hypotenuse = 0.014 in. (0.356 mm), and diameter (TP) equals radius times two or 0.028 in. (0.711 mm).

This 0.028 inch TP was often rounded to 0.030 in. (0.762 mm) TP by engineering—a nice round figure to work with.

Chapter 8

Temperature Control and Coolants

Temperature is a variable that must be considered on all large parts. Air-conditioned rooms have been used to improve dimensional control. Air-conditioned rooms cost a lot to build, equip, and maintain. During cutting or drilling, the cutter is usually flooded with coolant. Coolant temperature is a big factor in part temperature, but the machining community has largely ignored control of coolant temperature. The researcher found only one company out of eight visited that controlled coolant temperature. Electrical costs for air conditioning keep rising. It is also difficult to get people to work at 68°F (20°C). Workers tend to adjust controls for about 72°F (22.2°C) for personal comfort, despite management attempts to hold 68°F (my observations).

A second step is to improve part temperature control by controlling cutting fluid temperature. Temperature variations during the milling process lead to expansion or contraction of parts and of the drilling machine. Zelinski (1997) reports that investigations have shown that up to 80% of part errors are caused by thermal variations (Zelinski, P., Shop with a nervous system. <u>Modern Machine Shop</u>, 1997.). Temperature variations are caused by heat produced by milling, coolant, sun, air conditioning, and/or heating systems. The researcher considered temperature to be under control if variations are maintained within 2°F (1.1°C) on large parts. The coefficient of thermal expansion for the titanium parts and steel tooling most prevalent in this study is approximately 6.6×10^{-6} in./in./°F (11.9×10^{-6} mm/mm/°C). For an 80 in. (2.03 m) long part, this calculates out to be 0.001 in. (0.0254 mm) of change in length for each change of 2°F (1.1°C).

Statement of the Problem

Doors produced for some fighter aircraft are not the usual "hole covers," but are actually load carrying members as well as replaceable access doors. These functions require fastener hole locations produced to the closest possible true position (TP) accuracy. Engineers made the drawing requirements more precise based on predictions for improvement in machine accuracy and on design needs. The new specification called for a TP of 0.007 in. (0.178 mm), but the predicted progress to 0.007 inch TP had not been fully achieved by normal machining processes.

The accuracy of three large machines referred to as PDCs (Precision Drilling Centers) have been improved at considerable expense to meet precision needs. The resultant precision of these machines now surpasses that of normal five-axis NC machines by a considerable margin.

Projections of increased part volume for production indicated a lack of machine capacity on the three PDCs for production of selected parts. Additional machines were required to meet higher production rates. The needs for accurate hole patterns on certain large parts for the company's latest fighter aircraft exceed the precision capability of most existing machines in the world. Management decided to outsource the machining of these precision parts rather than purchase more capital equipment and to save floor space. This created the problem of not knowing if a given outside machine or process is capable of meeting the stringent 0.007 in. (0.178 mm) specification. The parts are too expensive to use as tryouts to determine a given machine's accuracy, so a method was needed for predetermining or anticipating the process capability of large five-axis drilling machines.

Drill Jig Accuracy and Limitations

The first process examined as a solution to our need for additional capacity was the use of drill jigs.

Rise and Fall of Drill Jigs

Precision hole patterns in large parts, especially on parts that are replaceable and interchangeable, have traditionally been produced using drill jigs. Some precision machines now equal or surpass the accuracy of drill jigs. Thus a movement away from the use of drill jigs began. Drill jig accuracy on hole positioning for large parts was measured and found to be 0.012 in. (3.05 mm) TP (True Position). Large milling and drilling machines were traditionally designed to produce part tolerances of ± 0.010 in. (0.254 mm) in each axis. This accuracy equates to about 0.028 in. (0.711 mm) TP.

The following elements in the life cycle of drill jigs are outlined and discussed:
- Conception and birth of the drill jig

Chapter 8

- Drill jig as a medium of inspection
- Temperature compensation
- Cost considerations
- Demise

On parts where 0.012 (0.305 mm) TP is acceptable, drill jigs will probably continue to be the process tool of choice. When precision higher than 0.012 TP is required, high precision milling and drilling machines will be used.

This report on the demise of drill jigs was written for inclusion in the appendix of my doctoral dissertation ***Predicting Accuracy of Five-axis Drilling Machines***. I was Chairman of the MTA/SME (Machining Technology Association of the Society of Manufacturing Engineers), 1999–2000 and have worked in the fabrication industry since 1953. The notion that drill jigs are becoming obsolete is strictly my opinion, and is based on the following reported information.

Accuracy and Robustness of Drill Jigs

Data on drill jig accuracy was ***not*** found in a 1994 search of literature and discussions with experts. Therefore, a drilling accuracy study was conducted specifically for this purpose[2]. The study was conducted on composite material and titanium alloys with several methods of drill jigs. The best dimensional repeatability for drilling titanium alloys was from lock strip indexing (a type of drill jig). Drill jig accuracy on hole positioning for large parts was 0.012 in. (0.305 mm) TP. This tolerance value is for plus or minus three-sigma tolerance limits, which means that three out of every 1000 holes drilled will be out-of-tolerance.

A further challenge to the figure of 0.012 in. (0.305 mm) TP of drill jigs was made in 1999. The researcher phoned six drill jig experts in various parts of the United States. To the question, "What TP can you hold with drill jigs on large parts?" the answers varied from 0.005 in. (0.127 mm) TP to 0.012 in. (0.305 mm) TP. All but one person stated that temper-

---

[2] This study was performed at the urging of Paul Hogle and others on the Boeing Gantry Team, and Terri Kalan in a Boeing Internal Report, "Drilling Accuracy Study Report," 1995, reported results.

ature must be controlled; most thought that materials for the drill jig must be compatible in coefficient of thermal expansion with that of the material being drilled. Once again, not one person knew of any specific data measurements on the results of using drill jigs.

Life Cycle of Drill Jigs

No attempt was made to pin down historical dates of the following events, but my observations propose reasons for the major steps in the evolution of the drill jig. The subsections are:
- Conception
- Media of inspection
- Temperature compensation
- Cost considerations (positive and negative)
- Demise

Conception and Birth of the Drill Jig

Drill jig accuracy on hole positioning for large parts is 0.012 in. (0.305 mm) TP. Milling and drilling machines were typically designed to hold 0.028 in. (0.711 mm) TP. Therefore, drill jigs were used on large parts when high precision was important.

Drill Jigs as a Media of Inspection

In their evolution, drill jigs began to accomplish more than just production of precision hole locations. In many instances, drill jigs were built accurate enough to allow them to be used as a media of inspection. This meant that hole patterns in parts drilled with a certified drill jig did not need to be inspected. The tool (drill jig) was usually qualified as a media of inspection by careful inspection and acceptance of the first part produced. This saved the cost and schedule time of inspection and reduced part cost.

Temperature Compensation

On large parts, temperature control is necessary because of the coefficient of thermal expansion of the parts and drill jig materials. One step in the evolution of drill jig accuracy was to use the tools in a temperature-controlled room. Another approach was to build the drill jig out of the same material as the part so expansion and contraction due to temperature would be the same in both. Sometimes a material with a similar CTE (coefficient of thermal expansion) was used for the tool. For example, the CTE of mild steel very nearly matches the CTE of most titanium alloys, so the less expensive steel is often used for drill jigs for titanium parts.

Cost Considerations

Positive considerations: Issues that favor the use of drill jigs are:
- The initial cost of a drill jig is high—perhaps $100,000 for a large part, but high-cost capital equipment is not required for the drilling operations during part production.
- Cycle time for large parts can be speeded up because more than one operator can be drilling at the same time.
- Negative considerations: Issues that are against the use of drill jigs are:
- Revisions to drill jigs are expensive and take time. This is detrimental when process improvements or design changes need to be accomplished.
- Costs for storage and retrieval of drill jigs are high.
- Size can become an issue on very large parts, say, parts over 12 ft. (3.7 m) long. Drill jigs get unwieldy and hard to handle if too big, and lack of rigidity becomes a problem.
- Maintenance costs for a drill jig are fairly high due to frequent damage, wear, and bushings spinning loose.

Demise of drill jigs

With the marriage of technology to precision machines, machines are now surpassing the accuracy of drill jigs. Technical improvements to NC (numerical controlled) machines in the form of precision scale or laser-positioning feedback have been instrumental in raising the precision level of large five-axis machines.

The crossover from drill jigs to machines happens when machines are able to produce holes to a positional accuracy that exceeds the accuracy of drill jigs. As noted above, drill jig accuracy on hole positioning for large parts is limited to 0.012 in. (0.305 mm) TP.

Maintaining Large Part Accuracy

Summary

On parts where 0.012 in. (0.305 mm) TP is acceptable, drill jigs will probably continue to be the process tool of choice because hourly costs are less and, on large parts, high priced capital equipment is not tied up, and the use of more than one operator reduces flow time.

Development of Precision Drilling Centers (PDCs)

Prior to the parent research study, there was a decision to build and use precision drilling machines for the drilling of large parts on a given program. This decision to abandon the use of traditional drill jigs followed many discussions on the merits of both methods. Precision hole patterns in large parts, especially on parts that are replaceable and interchangeable have traditionally been produced using drill jigs. Research done at the time of these discussions established the accuracy of drill jigs vs. that of precision machines. Some precision machines now equal or surpass the accuracy of drill jigs. Thus, a movement away from the use of drill jigs began.

Finding Other Precision Machines to Handle Production Overload

The objective of this data analysis is to produce control charts that summarize data from tested machines. These data should characterize the ability of the machine to accurately produce hole patterns in large contoured parts, and should allow a direct comparison of the accuracy of one machine to another. Every attempt was made to limit variations in hole positions to normal process variations. To minimize outside influences, all parts were drilled with drills furnished from the same purchased lot, all machines were furnished the same NC drilling program, and all parts were inspected with the same coordinate measuring machine (CMM) using the same inspection NC program and same inspection operator.

Development of a Test Fixture to Predict Machine Capability

In summary, the researcher developed a new re-usable test fixture and a process plan for testing large five-axis machines. A test method was designed to characterize the ability of a machine to accurately produce hole patterns in large contoured parts, and should allow a direct comparison of the accuracy of one machine vs. another.

The objective of this newly-developed statistical method for **predicting process capability** is the prediction of accuracy for five-axis drilling machines producing large parts. The researcher decided to calculate the "worst case" prediction of true position (TP) for each of the machines tested to confirm the credibility of this method.

Analysis of New Statistical Method (Research Task 2):

The researcher developed a new re-usable test fixture and a process plan for testing large five-axis machines. The new design provides six lots of four subgroups of data for completion of a statistical control chart. The new test method characterizes the ability of a machine to accurately produce hole patterns in large contoured parts, and allows a direct comparison of the accuracy of one machine vs. another. Data evaluation involves:
- Correction of X and Y values to projections to the XY plane using cosines.
- Performing adjustments in values based on temperature variance from 680F (200C) during drilling.
- Doing a best fit of TP data for each part.
- Plotting a histogram and control chart.

The measuring tool (CMM) has an accuracy of ± 0.001 in. (0.0254 mm) in X and in Y values (uses over one third of the tolerance specified for the part). Each of the 24 test parts being measured needs to be best fit. This involves a best fit of the five holes that constitute a given test part. Routines for determining best fit were written.

The calculation of TP requires squaring the X and Y values and results in a non-normal curve, but proof is given (from research) that samples of data from this non-normal curve do plot as a normal curve. Therefore, normal statistical analysis for control charts was used.

The objective of this newly-developed statistical method for predicting process capability is the prediction of accuracy for five-axis drilling machines producing large parts. The researcher decided to calculate the "worst case" prediction of true position (TP) for each of the machines tested to confirm the credibility of this method.

The approach taken was to shift the values for a normal curve for each machine tested by the value of the standard error of the mean. Two shifts in values were made. The first shift in values was 1.2% for the use of a smaller sample size (24 parts vs. 50 parts recommended), and the second shift was 7.2% for measurements of five holes in test parts vs. measurement of 150 holes in actual parts. The total shift is 8.4%. Calculations of standard error of the mean were made for a confidence level of 95% (two-sigma) for the twelve machines tested, and the calculated accuracy of the new approach is within 91.6% of the accuracy of the traditional statistical methods.

Large milling and drilling machines were traditionally designed to hold part tolerances of ± 0.010 in. (0.254 mm) in each axis. This accuracy equates to 0.028 in. (0.711 mm) TP. The 0.028 TP is a direct calculation from the ± 0.010 in. (0.254 mm) accuracy of machines for each of the X and Y-axes,

Results and Findings of Testing

The first machine tested was one of our PDCs. The test results from this proven machine became the benchmark of accuracy that other machines had to meet or beat in order to make the bidder's list.

The machines selected were touted by their owners to be extremely accurate. I will not name the machines tested due to company policy of not endorsing given products.

Seven of the machines tested did meet or exceed the benchmark accuracy of our current PDCs. The winning bidder has produced parts compliant with the 0.007 in. (0.178 mm) TP requirement for the past five years.

CHAPTER 9

MILLING DIFFICULT FEATURES

The purpose of this chapter is to present parameters for milling some difficult features in titanium alloys. The chapter begins with a review of efficient milling feeds, speeds, coolants, and avoidance of cutter burns, but primary focus is on milling thin webs and thin flanges in titanium alloys without the expense of back-up tooling, and on producing rough finishes in response to an unusual request. I have spent the past fourteen years developing processes for fabrication of defense and space parts from titanium alloy plates and castings. These processes were developed during the period 1991–1997, while resolving various milling problems on titanium alloy parts.

The order of reporting is:
1. Fundamental milling parameters for titanium alloys
2. Milling thin webs in titanium alloys
3. Milling thin flanges in titanium alloys
4. Producing a rough surface

Fundamental Milling Parameters for Titanium alloys

As a synopsis, when milling titanium alloys, sfm (surface feet per minute) are 60 sfm (18.3 m/min) maximum for cobalt (M42) cutters and 120 sfm (36.6 m/min) maximum for solid carbide (micro-grain or sub-micro-grain carbide) and carbide insert cutters. Chip loads are usually 0.005 in. (0.127 mm) and a maximum of 0.008

Chapter 9

in. (0.203 mm) per tooth for both types of cutters. This study on difficult sections was written prior to the breakthroughs in milling reported in chapters 6 and 7.

Coolants

We take a lot of care to use a heavy flood of coolant near the point of tooling contact. There are many good coolants available, look for:
- Better-than-average surface finish
- No bacteria problems
- Good cutter life and cutting action
- Environmental acceptance

Cutter Burns

Recent cutter burns in titanium alloys were due to cutter wear that suddenly accelerated as heat built up on the worn cutter. Burns are best prevented by:
- Using sharp cutters
- Flood of coolant
- Frequent examination of cutters for wear
- Paying attention to coolant action—liberal flow is a must
- Keeping chips away from the cutter to prevent re-cutting and interference with coolant

Cutter Life

Our goal for cutter life was a minimum of 45 minutes in these studies. The cutter life goal is usually met with 60 sfm (18.3 m/min.) maximum for cobalt cutters and 120 sfm (36.6 m/min.) maximum for solid carbide and carbide insert cutters. Chip loads were a maximum of 0.008 in. (0.203 mm) per tooth for both types of cutters. Cutters were considered dull when cutting edge measured 0.004–0.008 in. (0.102–0.203 mm) of wear or if the cutter became chipped.

I would recommend that any future cutter life testing be in accordance with International Standards ISO 3685 (tool-life testing with single-point turning tools), ISO 8688-1 (face-mill tool life testing, and ISO 8688-2 (end-milling tool life testing).

Milling Difficult Features

Milling Thin Webs in Titanium alloys

Minimum web thickness (floor of pocket thickness) in machined titanium alloy parts, such as aircraft spars; was traditionally limited to 0.100 in. (2.54 mm) unless expensive back-up tooling was fabricated to enable the normal tolerance of +/- 0.010 in. (0.254 mm) to be held on thinner pocket floors or unless chemical milling was used. The limitation of 0.100 in. (2.54 mm) minimum often blocked efforts to decrease aircraft weight because of higher fixturing costs.

Objective of Study

The objective of this study was to develop milling methods that would allow thinner webs to be machined to the required +/- 0.010 in. (0.254 mm) tolerance **without the need for back-up tooling**.

Accomplishments

Testing was conducted in two phases:
- A milling process was developed to produce webs that were 0.060 in. (1.524 mm) thick and within the tolerance of +/- 0.010 in. (0.254 mm).
- The process was further refined to produce webs as thin as 0.040 in. (1.016 mm) thick and within the required tolerances without the need for backup tooling.
- The developed process does not require back-up tooling or extended milling time and the required surface finish of Ra125 minch (0.0032 mm) or better is met. No oil canning of webs was evident. No tool chatter or unusual noises were noticed.

New Process

The sequence of operations, cutter motion, and cutting parameters used were as follows:

Sequence of operations:
    1. Rough to within 0.250–0.500 in. (6.35–12.7mm) of net thickness.

2. Measure thickness of part at this point if desired.

3. Finish mill to net thickness.

4. Cut off tool tabs (final operation).

- Cutter motion:
    1. Cutter ramped at 4 degrees from outside edge of pocket to center line—ramp angle depends on recommendations by cutter supplier for the geometry selected.
    2. Work from center of pocket to outside walls to produce net web thickness and fillet radius.
- Cutting parameters.

    RPF Cutter (Brand name)

    sfm = 150 (45.7 m/min)

    In./tooth = 0.004 in. (0.102 mm)

    Tool life history: = 45 minutes

Test Confirmation

A test part with five pockets was machined and measured to evaluate producibility. Web thickness varied from 0.039–0.046 in. (1.000–1.168 mm) with no oil canning.

Production Confirmations

We have produced over 100 production spars with pockets similar to the test part and have met the +/- 0.010 in. (0.254 mm) tolerance in all cases for the pocket floors.

Other Applications

This same process helps to minimize warpage in other thin sections such as thin splice plates. Even when the part is resting on a flat plate (which amounts to back-up tooling), we get good distortion control. Here too, the tooling tabs are cut off last.

Benefits

The new process has allowed designers to reduce weight by approximately one pound per one hundred square inches of web surface without the expense of back-up tooling. It applies in those instances where thinner webs are of sufficient strength.

Prologue

My thoughts on why this process works are that traditional milling spring passes do not work well on titanium alloys. Thin clean-up passes tend to push metal away from the cutter in the Z-axis direction and they impart heat and surface stresses into the part. The reason this process of leaving a heavy cut for last works is because the pull of the cutter helix balances the push of the cutter. One of Boeing's suppliers said he got best dimensional results on titanium alloys by leaving plenty of material for the final milling pass on pocket floors. Cutter dullness could certainly upset the balance of forces. By not doing final spring passes, you can actually save time and produce a dimensionally better part.

Cutting from the pocket center first to the outside wall last leaves plenty of solid/rigid material next to the cutter at all times.

Milling Thin Flanges in Titanium Alloys

Minimum flange thickness in machined titanium alloy parts, such as aircraft spars; was limited to a height to thickness ratio of 20:1 with a minimum of 0.060 in. (1.524 mm). To machine thinner flanges required expensive back-up tooling or chemical milling. These limitations often blocked efforts to decrease aircraft weight. Figure 9-1 shows these design limitation standards.

Chapter 9

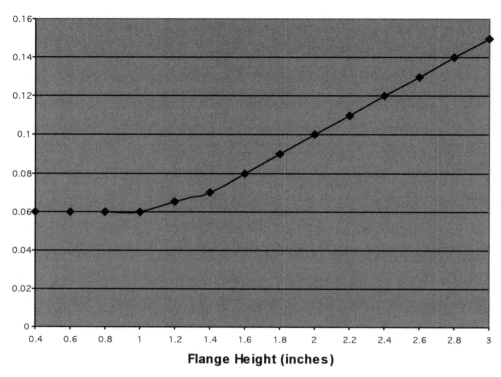

Figure 9-1: Minimum Flange Thickness vs. Height for Titanium 6AL-4V

Objective of Study

Develop milling methods that would allow thinner flanges to be machined to the required +/- 0.010 in. (0.254 mm) tolerance without the need for back-up tooling.

Testing

Tests confirmed the original design limitations. No breakthroughs here.

New Process

Convince engineering to add a note to the drawing that would allow machine mismatches along the sides of the flanges. Flanges as thin as 0.040 in. (1.016 mm) thick can be machined. The sequence of operations is:

Milling Difficult Features

1. Rough to within 0.5 in. (12.7 mm) of net thickness.
2. Finish mill the top 0.5 in. of the flange.
3. Step down in 'Z' and mill the next 0.5 in. of flange to net thickness.
4. Continue stepping until finished.

Production Confirmation

We have produced areas on many production parts with 0.040 in. (1.016 mm) thick flanges and have met the +/- 0.010 in. (0.254 mm) tolerance in all cases. The cutter mismatch is then hand blended/sanded if required.

Benefits

The new process does allow designers to reduce part weight. Fixturing costs are held low because back-up tooling is not required, and the milling cost is usually less because we have eliminated the final spring passes.

How to Produce a Rough Surface Finish

Sometimes it is necessary to produce a rough surface finish. We recently needed a rough finish of Ra 100–175 minch (0.0025–0.0044 mm) to better allow X-ray detection of voids caused if the electron weld beam wanders off the joint. The following worked:

- Cutter body, Carboloy brand, model P220.17-02.00 square shoulder, 2 in. (50.8 mm) diameter
- Use one insert in this three-insert cutter body; use false inserts in the other two positions for balance and stiffness.
- Use inserts Kennametal # (TPG322KC730) with 0.030 in. (0.762 mm) radius, ISO # (TPGN-16 03 OSE), Boeing A00 14650.
- Run spindle at 180 RPM.
- Use 2.9 in./min. (73.66 mm/min.) feed. It is critical that the feed rate does not decrease due to changes in rates while driving corners. If the feed rate falls outside the range of 2.4–3.1 in./min. (61–79 mm/min.), the

generated finish will not meet the required finish.
- Axial depth of cut 0.005 in. (0.127 mm).
- Tilt the head 0.1 degree in the direction of travel to keep the heel of the cutter from rubbing on the cut surface.

Prologue

The edges of the initial weld joints fit so well with normal milling of approximately Ra 63 minch (0.0016 mm) that we could not detect parts of the joint that were **not** welded by either X-ray nor by sonic methods.

Achieving a rough surface finish was not as easy as anticipated. All our experts remember rejections for rough finish and thought this would be easy. It proved to be very difficult because all of our efforts for over forty years have been aimed at improving surface finish.

Lifting the heel of the cutter was a key step because the trailing edge of the cutter tended to smooth out the roughness created by the front edge of the cutter.

While we were celebrating our success, the first production part went through the deburr department without good communications and was sanded smooth. We ended up having to re-do the rough finish on the part. The repair used about 0.005 in. (0.127 mm) of material.

Summary

Three major milling problems are presented with solutions. These problems are:
1. How to achieve thin sections in pocket floors that are 0.040 in. (1.016 mm) thick without using backup tooling.
2. How to achieve thin flanges.
3. How to achieve a rough finish of Ra150 minch (0.0038 mm). The solutions presented are efficient from a cost standpoint, and are proven over time with over 100 parts produced for each solution.

CHAPTER 10

WARPAGE

Straightening of titanium alloy parts and solutions to warpage are being covered in a separate book, but this chapter on warpage is included here for completeness as part of the milling process.

Problem

Large titanium alloy parts often warp beyond specified dimensions, and a lot of time is expended deciding whether to scrap the part, live with bad dimensions, or attempt to straighten the part. Warped parts cost money to fix or replace and cause schedules to slide. Unless the process is repaired, you will continue to produce bad parts.

Purpose

The purpose of this chapter is to help curb the losses caused by warpage. Discussion focuses on ways to identify the cause or source of warpage stresses and of repairing processes to reduce warpage on large titanium alloy parts. The proposed solutions to warpage are not perfect—there is still much to learn in this area—but the solutions discussed are tried and proven in production and are based on considerable research.

Chapter 10

Coverage and Limitations

This chapter deals primarily with warpage and emphasizes preventive measures. As stated above, the problem with warped parts is that they cost money to fix or replace and that warpage results in schedule slides. This book does not cover straightening of titanium alloys nor is it a primer on NC programming techniques. Details on straightening of titanium alloy parts will be covered in a future book as mentioned above.

The developing sections will try to explain the causes of warpage and ways to prevent warpage.

What Causes Warpage

Warpage for this discussion is caused by stresses within the part that are out-of-balance. Physically-damaged parts are not discussed.

Three Prime Sources of Stress

I have had to deal with many warped parts over the years, and have done considerable research of technical papers on warpage and stress.

My findings are that stresses can be grouped into three areas:
1. Inherited stresses
2. Clamped-in stresses
3. Machined-in stresses

Inherited Stresses

Inherited stresses are stresses that already exist in the raw material as purchased. Inherited stresses are caused by such practices as rapid cooling during thermal processing or cold straightening of the raw plate or forgings. These stresses show up when initial skin cuts are made, releasing some of the stresses.

An example of inherited stress happened a few years back when some stretcher-level aluminum plate began to warp and even break bolts in fixtures when milling the first surface. A study found that a well-intentioned employee in the shipping area of the material supplier was sending aluminum plate back into the shop to be straightened because it didn't look nice and flat. The straightened material we received looked beautiful, but the material (several million dollars worth) came from one supplier, and the material was all stretcher-level plate that is by definition supposed to be stress-free. Halting this flattening process solved the warpage problem.

A second example of inherited stress showed up in some ten-foot long forgings for titanium alloy spars that had been "stress relieved" prior to delivery. Once again a trip was made to the supplier and it appeared that cooling following stress relief was too fast. Rapid cooling in the final process of stress relief on this part with thick sections next to thin sections put stress back into the part. The forging supplier denied that they were cooling too fast and we went home. It is interesting to note that we have not had any of these parts warp in the four years since our visit to this company that was "doing nothing wrong."

A third example of inherited stress was when warpage occurred on a regular basis following an interim stress relief operation. Here the problem appeared to be caused by an undocumented cold straightening operation following stress relief. Many materials like aluminum and steel can be cold straightened, but titanium alloy parts cannot be cold straightened because cold bending changes the metal properties. In the case of titanium alloys, fatigue life goes to about half of its original value. Once again there was no admission of doing wrong, but the problem disappeared following our visit.

Clamped-in Stresses

Clamped-in stresses are best described by an example. Here a part does not fit down into the fixture until "big Charlie" stands on the part while the operator tightens the clamps. Later when the part is unclamped, and this often happens while a supervisor is standing by, the part springs out of the fixture and the operator exclaims, "I wonder why that happened."

Chapter 10

Machined-in Stresses

The heat of milling causes machined-in stresses. The third cause of warpage is stress due to milling. Here heat stress is put into the surface being machined and the result is a warped part. Dull cutters and lack of coolant exaggerate this stress by causing even higher heat. The main culprit is a dull cutter—I can't say enough about using sharp cutters and plenty of coolant to keep the heat down.

Short-Term Solutions

The obvious short-term solution to warpage is to save the part by straightening. As mentioned above, cold straightening below 1200°F (649°C) works fine for many metals, but cannot be used on titanium alloys when fatigue life is important. Notes on the drawing that allow an interim stress relief operation usually cover hot or creep straightening.

Support fixturing that will function at 1350°F (732°C) and weights are used to "creep straighten" parts made of ti-6Al-4V-ELI. This is usually done in a vacuum furnace to minimize the formation of alpha-case.

Shot Peening

A trial part (over wing fitting) about 3 ft. (1 m) long, forged from ti-6Al-2Sn-2Zr-2Cr-2Mo-STA was shot peened in a Wheelabrator on a rotating table. Typical part cross sections were 0.50 in. (12.7 mm). There was distortion up to 0.010 in. (0.254 mm).

Conclusions:

1) Ti-6Al-2Sn-2Zr-2Cr-2Mo-STA distorts from the stresses induced by shot peening.
2) Distortion can be controlled by process.
3) Shot peening can be used as a straightening or shape-altering process.

Warpage

Straightening

An aft boom side fairing was distorted up to 0.50 in. (12.7 mm) from nominal to 0.250 in. (6.35 mm) per side.

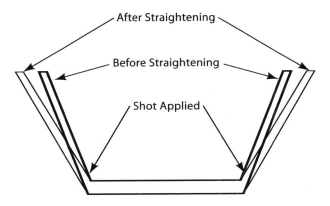

Figure 10-1: Straightening by Shot Peening

Straightening by shot peening corrected part to nominal +/- 0.050 in. (1.27 mm); see Figure 10.1.

A full discussion of straightening is not presented in this book.

Long-Term Solutions

Long-term solutions to warpage require investigation to determine the source category of the specific warpage stresses. These are the three prime areas of stress mentioned above.

The resolution of **inherited stresses** was implied in the examples presented earlier.

There are essentially two approaches to resolving **clamped-in stresses**. One solution is to force the part into the fixture and gamble that the stresses of milling on the opposite side will offset the original milling stresses and result in a straight part. The straight part will contain stresses that balance out to yield a straight part.

Chapter 10

Question? Have we just created a part with stress problems that will affect the life of the end product? This issue was taken up with our stress engineers, and they say this is not a likely problem because many of the parts tested for given vehicles were produced with similar stresses because they were machined in similar ways.

The other solution to **clamped-in stresses** is to shim the part so it is not forced into the fixture. This solution works if there is enough *cover* material to allow a full clean up of the part in final milling, and if additional milling warpage is not expected.

The long-range solution to **machined-in stresses** is often solved by changing the sequence of various roughing and finish milling media (flip-flopping the part in a different sequence).

Use sharp cutters and large volume of coolant to reduce part warpage.

Another solution to machined-in stresses is contained in the following discussion: Titanium alloy parts that have one flat side and pockets can have the pocket floors readily machined to thin sections without warpage or "oil canning" because the flat side is readily backed up with tooling. But with today's computer-aided design software, our engineers often put pockets on both sides of a part and want thin pocket floors to reduce weight. Up until eight years ago, we limited design folks to pocket floors that were at least 0.100 in. (2.54 mm) thick to eliminate the use of expensive back up tooling that would have to reach up into the pockets.

Research in two-phases has taught us how to machine pocket floors down to 0.060 in. (1.524 mm) and today down to 0.040 In. (1.016 mm) thick without back up tooling, and we do not get warpage or "oil canning". There are two tricks to doing this.
1. Cut the pocket floor from the center–out on all pocketing cuts,
2. On the next to last pocketing cut, leave 0.250–0.500 in. (6.35–12.7 mm) of pocket floor thickness for the final milling pass.
3. Make the last pocket floor milling pass to the final pocket floor thickness

Cutting from the center of the pocket out means that the cutter is always cutting next to rigid material. Radial depth of cut is generally 40–50% of cutter diameter. See *Chapter 9: Milling Difficult Features* for more details on this pocketing concept.

I surmise that the reason this heavy final pocket floor cut works is that the cutter helix pulls on the material and balances the normal push of the end of the cutter. The forces seem to balance out. Today we are cutting 8 x 8 in. (203 x 203 mm) pockets 0.040 in. (1.016 mm) thick with no warpage or "oil canning" and without the use of back-up tooling. Of course, with back up tooling one can machine down to foil thickness without warpage.

I found that this same technique solves another warpage problem. Splice plates that are flat on one side and have patterns on the other side tend to warp about 90% of the time. It appears that we get into warpage trouble by taking very little or no material off the flat side of the part. The answer here is to treat the flat surface of the splice plate as a pocket floor. If there is enough material thickness, take about 0.38 in. (9.7 mm) of material from the flat side and mill from the center out. We treat the flat side of the whole part just like a pocket floor and get no warpage.

To minimize warpage, try to mill the part out of the center of the stock—take an equal amount of material from each side of the stock. The short cut of only cleanup milling of the flat side of the part unbalances the stresses and causes warpage.

GROUP IV: SANDING AND VIBRATORY FINISHING

Sanding and Benching and Vibratory Finishing.
Achieving the final specified surface finish on titanium alloy parts is time consuming because of material hardness and specifications that are more demanding than for most other materials. *Chapter 11: Sanding and Benching* covers hand sanding and mechanically-assisted sanding. *Chapter 12: Vibratory Finishing* deals with mass media finishing that while commonly used for materials other than titanium alloys and on small parts, is just starting to be employed for large titanium alloy airframe parts.

Efforts to get specification relief will probably not succeed because sanding reduces the fatigue life of titanium alloys.

Studies to speed up the sanding process will not bear much fruit because the heat from fast or aggressive sanding also reduces the fatigue life of titanium alloys. There are also safety issues from the potential for titanium fines to catch fire, and titanium fires are difficult to put out.

The most likely area for reducing benching costs is to employ some means of vibratory-finishing.

In summary, the results of current and past research on the fatigue life of sanded titanium alloys requires that a Ra 32 minch (0.0008 mm) surface finish must be met if sanding is involved. The challenge is that we must develop a surface finishing method that does not reduce fatigue life, and that does not expand the cost of milling to astronomical levels. The probable answer to these costs is to introduce vibratory finishing, discussed in the next chapter.

CHAPTER 12

VIBRATORY FINISHING

The purpose of this chapter is to mention some early findings about the potential of using vibratory finishing (mass-media finishing) on titanium alloy parts. Vibratory finishing, as you know, has been used for years, especially on high-volume production parts made of many materials. Some vibratory finishing has been used on small titanium alloy parts because companies already own small vibrating equipment. Vibratory finishing of titanium alloys is now beginning to come into use, and is being used on some airframe parts in at least one overseas company.

This chapter will discuss:
- The notion that vibratory finishing does seem to be feasible.
- Some of the design features of successful vibratory finishing equipment.
- Media considerations.
- OSHA environmental issues.
- Part fixturing requirements.
- Where to get help.

Boeing specification allows vibratory finishing of titanium alloys.

Chapter 12

Vibratory Finishing is Used Overseas

A fellow research engineer did visit an overseas titanium alloy fabrication company. There he saw long titanium alloy parts (I call them one-dimensional) parts like spars being vibratory finished and vibratory peened. The equipment being used could be described as a dual shaft drive conventional vibratory finishing system. This type of vibratory system has been available from an American manufacturer since the late 1950s or early 1960s, but has fallen into relative disuse in recent years in this country.

Early Discussions With Machine Houses

The following points of discussion cover most of the elements of vibratory finishing, and are from meetings with machining suppliers of titanium alloys. The areas discussed included:

Vibratory deburring equipment:

Design time—per Miro Oryszczak of Giant Finishing (a Chicago finishing machine designer and builder), design time would be nil because all equipment discussed is already designed. There may be some equipment re-scale for size of parts, but no new designs.

- Build and delivery time—about five months.
- Foundations—none required, no isolation required for vibrations. A sound concrete floor 6 or more inches thick is required.
- Sound—the equipment generates 90+ decibels of sound. Equipment is normally covered with a lid (box like) that reduces the sound to 85 decibels or less.
- Part tubs and machine bases are integral—considered one-piece.

Part Fixturing

- Not all parts will require fixturing. Some parts will deburr and finish in a **free floating** mode.

- Some parts will need to be **rotated** on special fixtures within the tubs during deburring.
- Some parts will need to be uniquely fixtured by **clamping in fixed positions** with in the tub.
- Some fixtures can be built with adjustable ends so one fixture can handle a **family of parts**.

Media for Finishing

- Ceramic media will likely be the first step in the process.
- Some testing will need to be done to see if the Ra 32 minch (0.0008 mm) finish can be accomplished in a single media.
- Most companies use a second media of hardened steel to get a peened effect. Most of our drawings on titanium alloy parts for military aircraft do not require shot peened parts.
- If a second media is required, a second machine will usually be needed because of the high changeover time for changing from one media to another.

OSHA Considerations for Media and Liquids

Per Miro Oryszczak of Giant Finishing, the water, fines from ceramic media, and small amount of detergent used are not considered hazardous and do not require special hauling expenses. The water can be filtered and re-used in the process.

Support Team

Getting a vibro-finishing operation going on large titanium alloy parts will require a team effort and team support. It is recommended that the team would assist with:

- Machine selection for vibro-finishing
- Fixturing concepts and design
- Media selection
- Contacts—a network of specialists (see discussion below)

Chapter 12

- Process implementation
- Environmental issues, if any

Network of specialists—the Society of Manufacturing Engineers (SME, web address, www.sme.org) has a group of specialists that can be called upon for advice in this vibratory finishing area.

Summary

In summary:
- Drawings do not need to be revised to allow vibratory finishing because present drawing specifications already allow this process.
- Proven equipment designs for the necessary vibratory equipment do exist in this country.
- OSHA environmental issues are minor.
- Part fixturing will often be necessary.
- A network of advisors exists through SME.

GROUP V: CUTTERS

Sources of Cutters, Cutter Checking, and Cutter Life.

One really can't separate cutters from milling, but I have created a separate group for cutters because there are many specific cutter items to discuss. *Chapter 13: Sources of Cutters* presumes you have a cutter design or functional need in mind. It talks about the decision to make or buy the cutter. *Chapter 14: Cutter Checking* covers checking cutter dimensions on and off the machine and discusses some of the ways of detecting worn or damaged cutters. *Chapter 15: Cutter Life* attempts to discuss the effects of cutter material, coatings, feeds and speeds on cutter life relative to milling titanium alloys.

CHAPTER 13

SOURCES OF CUTTERS

Every machine shop has to decide whether to make cutters or buy them. (There is a similar decision to re-grind cutters in-house or to buy this service.) Another consideration is cutter coatings—what coatings to use, when to use coatings, and whether one re-coats after re-sharpening. Then there is the matter of tool holders (cutter holders.) What kind of tool holders are best for me? Another question is, "Do I need to balance the tool holder/cutter assembly?"

The purpose of this chapter is to discuss the many pros and cons that enter into these make vs. buy decisions. In this chapter, you will find discussions on make vs. buy of standard cutters, cutter re-sharpening, specialized cutters, cutter coatings, tool holders, and balancing.

Standard Cutters

Make vs. Buy of New Cutters

A small percentage of suppliers grind and produce most of their own cutting tools from solid carbide blanks. In addition to producing their own standard-type cutters, they find that many cutters must be specially designed for specific jobs. Off-the-shelf items are not always specialized enough and sometimes there is a lack of consistency among these purchased cutters.

Chapter 13

Another practice that does not work out well is to hire re-sharpening and setting of cutters in tool holders by a company that is many miles down the road. When the tool setting and re-sharpening services are far away, incidences of late delivery often more than offset the slight savings in re-sharpening costs. This long distance supply is a difficult situation to manage successfully. If your cutter volume is large and your re-sharpening service is far away, try convincing your supplier to open a sub-station across the street from your shop.

If you do decide to buy your cutters, buy from a known, reliable source. I have seen many cases where a tool salesman shows up with new "magic" cutters. The first cutters test well, we buy more and the results vary—some good and some bad. We get behind on parts and rush to buy higher quality cutters.

Make vs. Buy of Re-sharpening Services

In operation, tool life is targeted for a specific number of minutes based on the needs and culture of your shop, say sixty minutes, and then all cutters are re-sharpened. Shops that grind/make their own new cutters also re-sharpen the cutters because the same equipment can perform both tasks. Two shops that grind their own cutters claim that they are able to produce a more accurate cutter than those purchased. Re-sharpening is done with the same specialized cutter grinders that make the new cutters and the result is re-sharpened cutters that last as long as newly-made cutters. This is an interesting statement because most companies that re-sharpen cutters that were originally purchased, report less cutter life after re-sharpening than they get with new cutters. In one of our company machine shops, we purchased all new cutters, then re-sharpened them ourselves. The re-sharpened cutters had a shortened tool life of about 30% over new cutters. To maintain high quality we had to use new cutters for all finish cuts: re-sharpened cutters were relegated to roughing cuts.

Cutter Coatings

On titanium alloys, I do not recommend coatings when milling at less than 70 sfm (21.3 m/min). This slow milling speed applies mostly to rough cutting with cobalt cutters. The selection of what type of coating to use for faster (finishing) cuts is discussed in *Group III: Milling of Titanium alloys*. What I will mention here is that if it is important to use coatings on a new cutter, then it is equally important to use the same coating on cutters after re-sharpening. Re-coating presents a bit of a logistics problem because most shops are not capable of coating cutters. The usual solution is to overnight mail cutters to some company that gives you the price and service schedule that meets your needs. Perhaps we can convince our favorite re-coating supplier to open a coating sub-station across the street so we don't have to send our cutters across the country each time we re-sharpen. I have talked with two coating suppliers and they are considering moving to be close to the customers.

Specialized Cutters

Some relatively small diameter cutters require a long reach into a deep section of the part and the cutters flex too much to make efficient cuts. The usual solution is to use a tapered-shank cutter for more rigidity. The cost for tapered shank cutters is quite high. One low cost solution to this problem is to create a telescope of heat shrink tool holders to maintain needed rigidity to extend the length of cutters where a long reach is required (see Figure 13-1). The cost is kept low because standard heat-shrink tool holders are used and because only the tip of the tool needs to be replaced when worn.

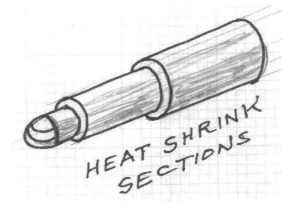

Figure 13-1: Telescoping Heat Shring Sections

Chapter 13

There are a host of other specialized cutters including key cutters, cutters with special shapes, and cutters with brazed-on wear edges. Without going into details on the many special cutters needed, let me just say that most special shaped cutters are made in-house. Some companies do hire the services of designing and building special cutters. The advantages of building in-house are short flow time and the ability to change one's mind and change the cutter design quickly as we go through several iterations of re-shaping to get the desired performance out of the cutter.

Tool Holders and Balancing

Many cutters in their tool holders are dynamically balanced before going to the mill. Balancing employs a system that drills balance holes or adds screws to ensure proper balance.

Some of the most successful shops use heat-shrink tool holders, and they are my choice for milling titanium alloys. Any cutter that runs beyond 8000 RPM must be balanced, but I like the notion of balancing tool holder/cutter assemblies for all RPMs because this act extends milling spindle life. It is also easy to manage the balancing process if all cutters are handled in the same manner.

One unbalanced cutter run at high-speed can quickly take-out a $50,000 spindle and shut a machine down for two weeks while waiting for a replacement spindle.

All balancing should be done dynamically using equipment that rotates the tool holder/cutter assembly. Static balancing is not adequate.

Summary

Reminders on cutters for titanium alloys:
- Make or buy quality cutters and keep them sharp.
- When using carbide cutters, re-coat after each sharpening to maintain maximum cutter life.
- Consider dynamic balancing of all cutters for longer spindle life.
- Use heat shrink tool holders.
- For long reaches, be frugal and make telescopes of heat shrink tool holders.

CHAPTER 14

CUTTER CHECKING

Cutter problems create scrap and rework. I have seen shops with scrap and rework rates that vary from 20% to 0.3%. When you have high scrap and rework, part quality and delivery are also affected. These are the consequences of cutter problems.

Quality and on-time delivery of machined parts are of extreme importance to the customer; quality and delivery are mentioned in one breath because they are almost inseparable. If a shop has quality problems, then on-time delivery does not happen. I suppose it would be possible to build quality products and then miss delivery deadlines through mishandling, but mishandling of parts is not discussed here.

An important step in producing good quality is having the right machines and equipment for the job. In an earlier chapter, we discussed rigidity, spindle run-out, tool holder run-out, machine bed accuracy, and coolant, but the most important step in producing good quality parts is use of the right cutter with the correct set length.

Proper cutter checking goes a long way toward eliminating these problems. Many parts are scrapped because of cutter problems. The primary cutter problems are:
- Installing the wrong cutter
- Wrong cutter material
- Wrong cutter diameter
- Wrong cutter radius

Chapter 14

- Having an improper set length (set length is the amount of cutter sticking out of the tool holder)
- Failure to detect cutter wear or breakage

The basic message of this chapter is to highlight the importance of sound cutter checking practices. From this chapter, you should learn methods for checking cutter accuracy and ideas on monitoring cutter wear.

The main body of this chapter contains a listing and description of cutter checking methods and ideas for monitoring cutter wear.

List of Cutter Checking Practices

Here are four practices that I see being used to check cutters, along with my comments. I list these practices in order of least effective to most effective (my opinion):

1. Having a second pair of eyes check the cutter and the setup in the tool holder.
2. Having an electronic chip mounted on the tool/holder setup that is looked at with a second pair of eyes and that is electronically verified at the milling machine. This becomes especially important if a cutter breaks or is lost and the operator hunts up a replacement. The electronic check must still pass.
3. Here we have a clever solution—the use of fail-safe pins. A fail-safe pin is a frangible pin located on the milling fixture. A sub-routine in the NC program is called on each time a cutter is changed. The spindle is turned on and the cutter is passed around the pin, diagonally up the pin, and across the top of the pin. The gap between the cutter and pin is about 0.002 in. (.0508 mm..) If the cutter diameter is too large, it hits the pin as it passes around it. If the cutter radius is too sharp, it hits the pin during the diagonal pass. If the set length of the cutter is too long, the cutter hits the top of the pin. If the cutter clears the pin during this

check, the program is allowed to continue because the cutter will not encroach on any material that should not be removed. If the cutter is incorrect by being too small in diameter, too large in corner radius, or if set length is too short, the part is not undercut, hence it fails safe. A variation of the clearance around a fail-safe pin is to use an aluminum pin that is lightly cut by the cutter, and replaced after each use (see Figure 14-1).

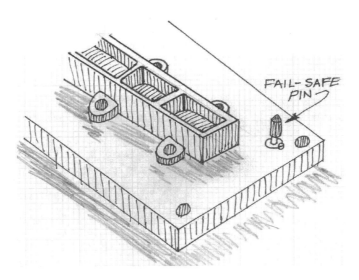

Figure 14-1: Fail-Safe Pin on Fixture

4. Install a laser tool checker such as Renishaw or Blum brands that check cutter diameter, cutter radius, broken teeth, and set length. The spindle is turning during this check. Here the set length and cutter radius offsets can be electronically altered on the fly. As mentioned above, this check will also detect a broken tooth.

In one shop I saw a completely improper use of fail-safe pins. Here the radial gap between fail-safe pin and cutter was 0.010 in. (0.254 mm). With this large gap, we actually had a cutter that was too large in diameter pass the fail-safe pin test and

Chapter 14

scrap the part because the machined walls were too thin. The gap must not exceed 0.003 in. (0.076 mm) to be effective; 0.002 in. (0.051 mm) of gap is better.

The impact of having the right cutter accuracy is huge. One shop that I worked with was running 20% extra costs due to scrap and rework without the use of fail-safe pins. The phrase *on-time-delivery* was not even in their vocabulary. This scrap and rework cost reverted to about 2% after fail-safe pins were installed.

I visited another shop that has eight mills, all with laser tool checkers integrated into their systems. Their scrap and rework has averaged less than 0.3% for over eight years now and their deliveries are always on time.

Ideas for Monitoring Cutter Wear

The problem of detecting heavy cutter wear or broken cutter elements is much more difficult than ensuring that the right cutter is being used. This problem may never be perfectly solved.

Here is a hodge-podge of thoughts in no particular order:
- Careful vigilance by the operator every few minutes by eye and by ear. This need for the operator to check does not work well when flooded pockets block one's view. This approach also does not allow one operator to run several machines or to be doing other productive operations while the machine runs.
- Installation of special audio equipment that listens to the cutter sounds and compares this to a known sound pattern for the cut being made. The machine is stopped when limits are exceeded.
- Horsepower monitoring with limits being set that will halt the machine.
- A variation of horsepower monitoring is to monitor variations in spindle horsepower between spindles on a multi-spindle machine where identical cuts are being made.
- Measuring the torque that drives the cutter is a great indicator of cutter wear. When milling titanium alloys our lab considers a cutter worn out when

the torque doubles—this equates to the cutter land being worn to about 0.005 in. (0.127 mm).
- NC programs can be written to call for a cutter change that tells the operator when to change the cutter even if it does not appear to be worn. These changes are based on known cutter life testing data. This approach also allows many cutter re-sharpenings, which keeps cutter costs down. I find that no two operators agree on when a cutter is worn and should be replaced. This decision is best controlled by the programmer.

Summary

Scrap and rework can be reduced to acceptable levels (under 2%) if good cutter checking practices are used. Fail-safe pins and laser cutter checkers are the best cutter checking methods that I have seen to date. Cutter wear also enters into the scrap and rework picture, so care in monitoring for cutter wear is a must.

CHAPTER 15

CUTTER LIFE

Cutter life is a major factor in determining the cost of milling titanium alloys—perhaps **the major factor**. The answers to the questions surrounding the variables that affect calculations for optimum machining costs are different for each company; there is not one clear answer that you can grab and run with. Each company has its own culture, but the essence of the variables that affect cutter life when milling titanium alloys are the same for every company.

The basic message of this chapter is that your aggressiveness with cutters affects cutter life, and, hence, cutter costs, and that same aggressiveness affects milling hours. The interplay of these two elements determines overall part cost.

This chapter is a discussion of the major variables that affect cutter life and the role that aggressiveness in cutting affects milling time. The key variables are cutter material, cutter coatings, and aggressiveness of milling. Cutter geometry is another variable, but will not be discussed in this chapter because 1) it is well worked out by the major tooling companies around the globe and 2) because I am not a great expert at designing cutters.

My hope is that you walk away better prepared to calculate the optimum solutions for your milling costs through discussion of the major variables mentioned above. These calculations will vary from company to company because of your company mission, labor rates in your shop, your decision to make or buy cutters, re-sharpening services, and your overhead and equipment costs.

General Notes on Cutters

In one of our major shops cutter life was required to be a minimum of 45 minutes when cutters were tested at 60 sfm (18.3 m/min.) maximum for cobalt (M 42) cutters and 120 sfm (36.6 m/min.) maximum for solid carbide (micro-grain or sub-micro-grain carbide) or carbide inserted cutters. Chip loads were a maximum of 0.008 in. (0.203 mm), but varied down to 0.002 in. (0.051 mm) per tooth for both types of cutters. The average chip load is 0.005 in. (0.127 mm) per tooth. Cutters were considered dull when the normally sharp cutting edge or land measured 0.005 in. (0.127 mm) wide due to wear or if the cutter became chipped. If you have the ability to measure cutter torque you will see the cutting torque approximately double when wear reaches 0.005 in. wide at the cutting edge.

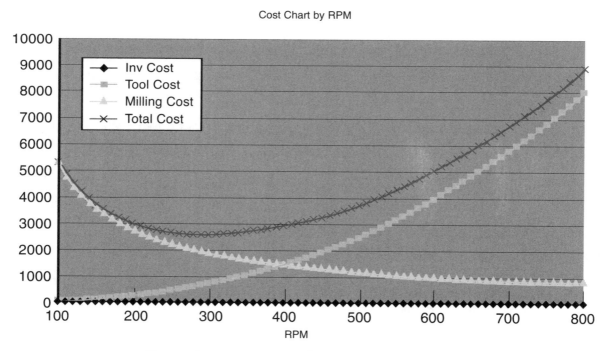

Figure 15-1: Minimized Rough Milling Costs

Cutter Life

My tests and experiences over the years point to high temperature as the main enemy of cutter life. Once a cutter becomes worn, the temperature builds up very fast, and we get meltdown in a matter of seconds. Sharp cutters and plenty of coolant is the remedy.

Recent cutter burns in titanium alloys were due to cutter wear that suddenly accelerated as heat built up on the worn cutter edges. Using sharp cutters and plenty of flood coolant best prevents the burns. Check the following list:

- A frequent examination of cutters for wear is in order.
- Attention to the coolant action to assure a liberal flow is a must.
- Chips need to be kept away from the cutter to prevent re-cutting and interference with coolant. My studies show a reduction of cutter life by as much as 35% when re-cutting chips in pockets during vertical milling.

Example of Minimizing Rough Milling Costs

Figure 15-1 depicts the total cost curve for rough milling of ti-6Al-4V-ELI on part X at varying RPM. These base parameters were used:

- Solid cobalt cutter (M42), uncoated—0.500 in. (12.7 mm) diameter, 4 flute
- Radial depth of cut = 0.20 in. (5.08 mm)
- Axial depth of cut = 0.50 in. (12.7 mm)
- Cutter cost $60.00 new and re-sharpening cost $11.00 each
- Chipload = 0.004 in. (0.1016 mm) per tooth
- Program efficiency = 70% (cutting air 30% of the time)
- Part setup time = 1 hour
- Manual tool change time = 0.1 hour

Optimum Results as Charted

- 290 RPM
- 76 sfm (23.2 m/min.)

Chapter 15

- 0.55 hours cutter life
- 8.7 in./min. (221 mm/min.) = feed rate
- 1.91 in.^{3}/min. (0.03 dm^3/min.) of titanium alloy removed

Note: Cost of carrying inventory is plotted, but has almost no effect on the optimum results.

Parameters for Finish Milling With Solid Carbide Cutters

In this subsection I list the standard parameters for finish milling and basic values for cutter life relative to sfm.

Standard Parameters for Finish Milling

The following parameters for finish milling with TiAlN coated solid carbide cutters will give **low cutter distortion and minimize chatter**:

Use 0.025–0.030 in. (0.635–0.762 mm) of radial engagement for finish milling cuts in titanium alloys

Use a maximum of 100% of cutter diameter as the figure for axial depth of cut to minimize cutter distortion.

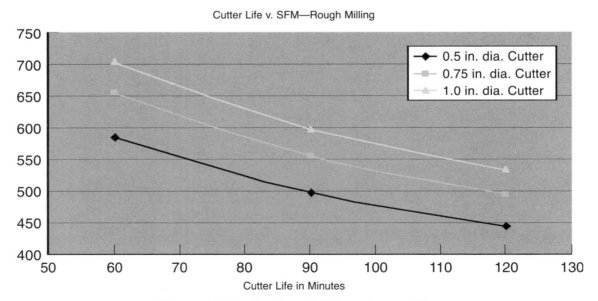

Figure 15-2: Finished Milling Cutter Life

Basic Values for Cutter Life vs. Sfm

The basic cutter life values for finish milling are a bit more complex than those for rough milling (see Figure 15-2).

Figure 15-3 shows the same results in metric values:

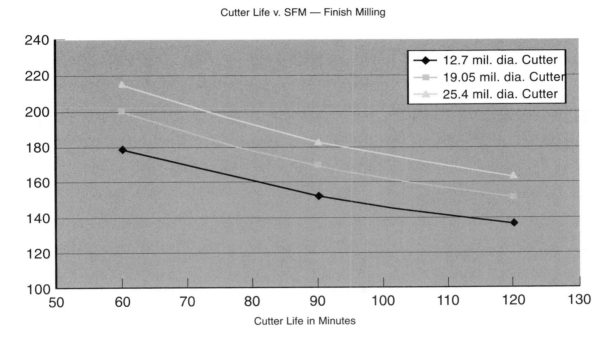

Figure 15-3: Finished Milling Cutter Life (Metric Values)

Finish Milling Costs

I did not include a chart for optimum cost for finish milling, but one could use a similar approach to the chart shown for rough milling.

Summary

Milling parameters for rough milling are reported for 60 sfm (18.3 m/min.) with uncoated cobalt cutters. One can expect longer cutter life at slower sfm—as much life as 8 hours at 30 sfm (9.2 m/min.), but I certainly do not recommend such inefficiency. For finish milling the recommendation is to leave only about 0.025 in. (0.635 mm) for radial cleanup and run cutters at an RPM that will approximate 600 sfm (183 m/min). Running at these parameters will often cut shop costs by 50%, but you need to include the values that work for your shop such as your dollar wrap rate.

GROUP VI: EFFICIENCY AND LEAN PRACTICES

Inter-Company Sharing, Continuous Improvement is Not Enough, Lean Studies and Communication, Metrics for Determining Shop Health, and Work Cells.

The objective of this group of chapters is to get costs and flow time down while maintaining high quality. *Chapter 16: Inter-Company Sharing* discusses the merits of sharing technology and ideas between companies. *Chapter 17: Continuous Improvement* points out that the norms we are using may not be enough. *Chapter 18: Lean Studies and Communication* is a key way of making quantum leaps of improvement in costs and flow time without sacrificing quality. *Chapter 19: Metrics for Determining Shop Health* is a discussion of a frugal way of measuring shop efficiency and why the more sophisticated methods usually fail. *Chapter 20: Work Cells* provides some history lessons in using work cells and the strong successes of today when tied to lean studies, and the importance of being flexible—being able to rapidly relocate and rearrange equipment.

CHAPTER 16

INTER-COMPANY SHARING

Professor Chris Brown, Ph.D., of Worcester Polytechnic Institute cautions;
Lean principles and cost savings are important. However, if all a company does is cut costs, eventually it will be out of business. Applying lean-manufacturing principles cannot replace product innovation and the value enhancement that results, nor can they produce significant process innovations and the potentially large cost savings available there. American manufacturing industry needs to invest in innovation and the things that support it, like education, research, and collaboration.

May I suggest that becoming ever more competitive, using lean practices may not be enough for survival in this world, and like it or not, we are now a part of the global world. Survival may require a sharing of technology between companies of like skills. The thoughts on sharing apply to any group of companies with like services. Companies might consider sharing technology to solve problems faster and better than would happen if a given company works at this alone. In one major sharing group of machining companies, progress has been made in many areas at a much faster pace than would have happened if inter-company sharing had not been occur-

Chapter 16

ring. Some areas affected are: speed of milling, reduction in the number of fixtures, reduction in setup time, efficient cutter design, cutter coatings, and influence of machine-design by manufacturers of equipment.

My earlier paper on this topic was entitled, *"Strengthening Your Company Beyond Lean."* Lean is a hot word right now and grand progress in efficiency is being made. For example, a Management Forum, *"Making It In America: Rising Up To The Global Manufacturing Challenge,"* was held starting June 10, 2004, alongside of the SME annual meeting in Cincinnati, Ohio. The forum focused on how small- and medium-sized manufacturers can become more globally competitive by applying the principles of lean manufacturing and continuous improvement throughout their enterprise. In addition, speakers discussed workforce training and development programs that will help companies streamline their manufacturing processes and accelerate speed-to-market of new products. The event also featured case studies by manufacturing experts who shared their firsthand insights on how they successfully reduced their company's operational costs while increasing efficiency, compressing product development cycles, improving productivity, and enhancing product quality—without having to resort to outsourcing production overseas, particularly to low-wage China.

Problem

All these lean steps may not be enough to keep your shop solvent. There is another step that can be taken that involves a major cultural adjustment—the idea of sharing technology and aligning yourself with your close competitors.

The purpose of this chapter is to:
- Show why lean is not enough
- Present examples of success from an eight-year experience in working and sharing of technical ideas
- Allay fears about giving away ideas and sharing technical information
- Additional thoughts

Inter-Company Sharing

Why Lean is Not Enough

Races do not have to be won by a lap or even 50 ft. (15 m). Races are often won by part of an inch. These are tough times, and my guess is that about 1/3 of the fabrication businesses will fail in the next two or three years. If most of your competitors are also working hard on lean studies, what will you do to give your company some edge that will mean survival?

Examples of Success From Sharing

Eight years ago, under Boeing's leadership, a half dozen companies formed a sharing group or team and called this a **Symposium**—which implies a learning experience. **Consortium** was considered too strong a term and has other linking aspects and possible legal ties that are not desirable. Boeing looks at our suppliers as an "extension" of the company.

It appears that lean studies are absolutely necessary and are the core of great improvements these days. But a few companies that formed a symposium have taken progress a step further. The most effective lean improvements in machining suppliers of titanium alloy parts that I have seen have happened within this symposium. Here, several competing companies actually conducted a "round robin" of lean studies. Each company, in turn, hosts a study with representatives from the other companies working and sharing together. The host company also brings in technology experts and speakers and everyone shares further as they socialize and break bread together. Every three or four months, a different company hosts a get together.

The members of the symposium are not identical. They are of different sizes and selectively adopt or develop shared ideas that best fit their business and products. Here are some of the areas of grand progress:
- Designing, making, and re-sharpening of your own cutters.
- Optimum use of cutter coatings for specific milling operations.
- Use of fewer fixtures—the challenge is one fixture—or even better, universal fixtures.

Chapter 16

- Cutter feeds and speeds for milling titanium alloys that match or top the best that I have seen in the industry.
- NC Programming techniques and standardizations that are setting trends in the industry.
- A sharing of ideas on reducing machine down time.
- Group pressure on machinery builders in this country and abroad to improve machine designs—not just waiting to see what machine manufacturers will come up with.
- Ideas on forming management teams where members are hired based on anticipated teaming characteristics: "We must work together as a team."

The success from this first symposium has inspired us to launch a second symposium of machining suppliers. This second symposium consists of seven machining companies. After one year, three major areas of improvement are evident from this second symposium:

1. The launching of preventive maintenance training—we call this PMT. This was a collaboration of representatives from the seven companies working for one-week in one company, and is expected to spread throughout the group.
2. I gave a seminar on plunge roughing of titanium alloys in May of 2004 and a report back to the new symposium by one member of the group (with a movie) showed a savings of 66% on a particular titanium alloy roughing operation that used to take about 4.5 hours per part.
3. A three-day lean study by representatives from six competing companies resulted in a cost avoidance of $450,000 in capital. The new approach was developed at a projected cost of $2000.

What is the proper number of companies to make up such a symposium? Our thinking on deciding how many companies to invite to a symposium was based on a projection of meeting three times a year and of having a given company host the event once each two or three years. So the desired membership is six to nine companies.

Inter-Company Sharing

Another area that is under consideration is to team up and share between members of the supply chain that tie to these machining houses such as: chemical houses, thermal processors, and inspection services.

Allay Fears of Giving Away Ideas and Sharing

I have seen a very healthy attitude in some companies that somewhat parallels Henry Ford's early principles that our task is to continue to be creative and stay a jump or two ahead of the competition. Then when they steal one of our ideas, we are at a new level of processing, a step or two ahead of them. No matter how hard we try, our competition does find out our good ideas. It's just a matter of time.

Summary

Performing lean studies may not yield enough progress for you to win the competition battle.

Lean practices plus sharing of technology and ideas between competitive suppliers can make your company even stronger. This strengthening through sharing may be necessary for survival

Grand progress has been made in two symposiums of machining suppliers that I work with.

CHAPTER 17

CONTINUOUS IMPROVEMENT IS NOT ENOUGH

The victory cry of Aesop's tortoise, "Slow and steady wins the race," strikes me as only rarely applicable, and surely secondary to the usual mode of human enlightenment, either attitudinal or intellectual: that is, *not* by global creep forward, inch by subsequent inch, but rather in rushes or whooshes, usually following the removal of some impediment, or the discovery of some facilitating device, either ideological or technological.

Stephen Jay Gould (2000) *Lying Stones of Marrakech*

Gould goes on to give two examples of rapid progress: the telescope and moon rocks returned by Apollo's Armstrong.

Outsourcing of machining by large companies is certainly the trend, but volume of machining work is not high enough for all suppliers to survive. While continuous improvement was adequate for keeping your shop busy in recent years, it probably does not provide enough impetus for growth in today's and tomorrow's markets. Major improvements in efficiency are required.

We better not take it for granted that big companies will continue to perform machining research and development, and dispense the findings to suppliers. It may not be long before large companies complete the task of outsourcing all machining. It will follow that research budgets for machining will also be eliminated. The task of improvement will then fall totally on the shoulders of machine shops and cutting tool suppliers.

Below I have reported on potential areas for rapid improvement, examples of sharp advances that have been made in milling, and a summary.

Potential Areas for Rapid Improvement

The sharing of technology between competitors mentioned earlier in this book could be a partial answer to the needs for rapid progress. In most of the areas discussed below, there are current technical papers available from the Society of Manufacturing Engineers and myself.

Cutting Tools

When does it make sense to sharpen my own cutters? Could I save money by making my own cutters? What about coatings? What about progress that is happening in the world?

Fixturing and Job Setup

Some companies are making grand progress in the area of fixturing by reducing setup time. Going to single fixtures and making use of universal fixtures are good steps. I recently saw a three-fixture job changed to a single rotisserie. This reduced the setup time from six hours to about 15 minutes. When reprogramming was finished, a 20-hour task becomes less than two hours on this production part.

Continuous Improvement

High-Speed Milling

You are most likely involved in changes in milling speeds on aluminum, titanium alloys, and other materials—but are you up to date? Progress is being made on titanium alloys at a rapid pace.

Part Warpage—Prevention and Straightening

One traditional solution has been to increase the number of part flip-flops and sequence of same to minimize part warpage, but there are other solutions.

Coolant Ideas

With current environmental pressures, control of coolant is a big area of concern. The coolant suppliers can provide a wealth of information here. To minimize environmental problems, consider the use of spray mists made up of mostly air and water for processes such as drilling of titanium alloys.

How about another idea—control of **coolant** temperature can be a "poor man's" answer to air-conditioning. It costs very little to take charge of coolant temperature and hot coolant—the fire hose wins the battle of part temperature over air conditioning every time. You do need to allow for some cooling by evaporation. I have measured temperature drops as great as 4^0 F (2.2^0C) in the part due to evaporation.

Benching

The introduction of vibro-finishing and deburring of parts usually yields the specified surface finish with very little labor. Because this is a controlled automated process, you will get improved quality because of the consistency. No two benching persons hand deburr and sand with the same results.

Examples of Rapid Advances in Milling

Here are some examples of advances beyond the normal improvement curve:

Finish Milling

Leave less material for cleanup in finish cuts (0.025 in. [0.635 mm] vs. 0.100

Chapter 17

in. [2.54 mm]) and use coated solid carbide cutters. This change can lead to an improvement of 8x in the finish milling speed of titanium alloys. Finish milling speed had not improved much in the last 30 years. Along with this change was the use of TiAlN coatings on the solid carbide cutters that doubled the cutter life from 45 minutes to 90 minutes and made the change very acceptable (see details in Chapter 7). The best example of high speed milling of titanium alloys that I have seen saved 400 hours on a difficult wing spar, and there is a left-hand part as well. This improvement involved use of a more rigid machine, fewer fixtures, solid carbide coated cutters, and fresh NC programming.

Plunge Roughing

The introduction of plunge roughing with four-inch diameter cutters allowed the use of carbide inserts instead of the former roughing with cobalt cutters (see details in Chapter 6). This change gives an improvement of at least 30% in overall roughing time of ti-6Al-4V-ELI, and a 4x improvement in specific cuts. The old rate was 5 in.^{3}/min. (0.08 dm^3/min.) vs. new rate of 20 in.^{3}/min. (0.33 dm^3/min). One roughing operation went from 4.5 hours per part to 1.5 hours per part in the best example I have seen to date.

Rotisserie Fixtures

I saw an aluminum part go from 20 hours floor-to-floor time to less than two hours floor-to-floor time. Here one fixture took the place of three and higher speed NC programming was employed at the same time. See Chapter 4 for additional fixture breakthroughs.

Milling Cutters

Two companies that I work with are now grinding their own solid carbide cutters for an average savings of 33%. Quality of cutters is claimed to be better. Re-grinds rival new cutters for tool life. The cutters are sent for coating after each grind.

Deburring and Finishing

The use of vibro-finishing equipment for titanium alloys is just getting started in this country. The savings here will eventually reduce the cost of finish milling because larger stepovers will be possible between finish contour cuts. I predict the savings will be several million dollars per year on major aircraft programs that make heavy use of titanium alloys.

Summary

My challenge to you is to seek ways to make major improvements—look for leaps and whooshes. Remember: the hare usually wins the race.

CHAPTER 18

LEAN STUDIES AND COMMUNICATION

Lean studies are the most efficient and effective way of making quantum leaps of improvement in your costs and flow time. They also help satisfy your customer who looks for the words "lean studies" in your bids and proposals. Your customers are looking for suppliers that practice lean because they want costs and flow time to come down as fast as possible. The customer asks in his mind, "Will the cost be less on future lots?" The basic message here is that lean efforts are probably the most efficient way to lower costs and flow time, and your survival depends on these kinds of results.

The purpose of this chapter is to outline the importance and effects of having sound lean programs. I do not discuss how to train for and conduct lean studies. There are a host of resources available for specifics on how to conduct lean studies. Lean training usually costs money, but do not overlook the opportunity for free training. Some major customers—large companies—offer free resources for training your people in lean practices as suppliers. The large customers want the advantages of lean practices enough to give this free training.

Some Examples of Quantum Savings

Here are some examples of lean-type studies that resulted in quantum leaps in progress. I had the good fortune to lead the following studies:

Chapter 18

- We had 15 full-time employees checking new tool design drawings before submission to NASA. We talked to NASA and asked them what would be the penalty for submitting un-checked drawings to them. NASA responded that they checked all incoming tool designs with a crew of 15 persons, and that there would be no contract penalty for errors (typo's and the like) getting through to them. So we got rid of the entire checking operation on our side except for one person (Management insisted we keep the one person in place.) By the way, the one person quit the company after six months because he was bored.
- In a similar study, we had 43 quality auditors keeping and maintaining records on the number of cycles that critical lamps and switches were exercised so we could replace them before they failed. For example, if a switch was known to fail at about 2400 cycles, we replaced it at 2000 cycles. I visited each of the seven recipients of the weekly reports we were generating, told them we were trying to reduce costs, and asked them if they could possibly get by without their copy of this report. Each of the seven managers signed a paper that they did not need the report. We went to our company lawyers and asked them to review our contract to see if the reports were required. Their answer was, "No." All we were required to do was keep enough records to tell us when to replace items before failure. Our solution was to hang a tab card at each switch. Employees kept score like the marks on a prison wall and when the number reached the maximum number of cycles, the switch or bulb was replaced. Now the only users of the reports were the 43 persons updating them for next week. The entire department was eliminated.

There were many other dramatic improvements. The point is that major improvements can be made.

Lean Studies

Process Stabilization

What constitutes a stable process and why is it important to wait on lean studies until a given process is stable?

We have also made some mistakes in trying to do lean studies. Sometimes we get anxious for savings and study a process that is not stable or has not jelled into a steady routine. The problem here was that we spent a lot of effort twice because we had to redo our studies after the process matured and stabilized.

Many Companies Do Not Use Lean

I am puzzled that more companies do not have solid lean programs. At a large SME conference in Cincinnati in mid 2004, a major lean consultant stated that only about 5% of the companies in the USA practice lean, and that half of the 5% are foreign owned companies building their products in the United States.

Summary

Our customers and we want ongoing lean studies because they reduce costs and flow time without degrading quality. In fact, lean studies often improve quality.

Remember to stabilize a process before doing a lean study.

Despite the potential for large savings, most companies do not conduct lean studies.

CHAPTER 19

MEASURING SHOP PERFORMANCE

A world-class shop knows the effectiveness and efficiency of their shop, and is able to measure progress at least quarterly. Some of the reasons why it is important to know your shop performance are:

- Accurate determination of when a job will be completed or when it will complete or start a given process step.
- Knowing which parts are making money for us and which are losers tells us which jobs to study for improvements.
- A record of how much improvement was realized from a given study (lean study).
- Knowing if the shop is making proper progress from period-to-period.

The most effective measurements or metrics are those that can be done quickly, timely, and efficiently. In this chapter, I will discuss some of the many systems used to measure shop performance, point out the shortfalls of complex systems of measurement, and elaborate on the merits of an ancient tool called the delay study, and discuss the efficiency of performing delay studies.

Timeliness is important. One shop I worked with recorded performance carefully, but only saw the results annually in March of the following year. The results that accounting came up with went something like this: Last year we spent an average of

Chapter 19

200% of hours against the standard work hours that we have on file for the many parts that were machined. Standard hours are the NC program optimum time from the computer plus estimates for deburring and other non-NC processes. So future bidding amounts to doubling the standard cost estimates for the next year until a fresh report comes out. That shop went out of business in 1999.

Some shops never record performance, and I wonder how they can give status reports that properly predict when a part will be ready to ship? How do they know which parts make or lose money? How do they know what parts or processes to study to make the most progress?

How is performance measured?

Some shops have employees record their time against the job they are currently working on. This is usually done via computer located at or near the employee's workstation. Data is wanded-in, bar-coded, or typed-in to the computer system. If your present system gives you all the data you need in a timely manner, then you could skip the remainder of this chapter.

Many of the current cost collection systems are complex and cumbersome because they require a lot of training and strong disciplines to make them work. The problem with some of these complex systems is that they just don't work. Or they work for a year or two, then fall into problems of lack of proper inputs. They are so complicated that they are hard to manage. These systems are also expensive to buy or develop the software, to administrate the system, and to pay for the operator's recording time on a daily basis.

My Recommendations for Performance Measurement

If your present system is not working properly, I recommend that you use old-fashioned delay studies to measure shop performance. Delay studies are inexpensive to perform and give results the next day with 90–95% accuracy.

A delay study is performed as follows:
1. Select which machines will be studied. If you have a small shop with less

Measuring Shop Performance

than a dozen machines, select all of the machines. In a large shop select a different group of about 10 machines for each shift. Select them randomly—use a table of random numbers or select numbers out of a hat. The randomness is important in a large shop, and selects a fresh batch of machines for tomorrow's shifts.

2. Select about ten random times of day for shop trips. I allow about ten minutes for a pass through the shop, so I break up the shop day into ten-minute increments omitting lunch break. Do not omit the first or last ten minutes from a given shift from the hat, but let random picks happen. This pattern will yield 100 readings per shift. Do not omit any shifts. If you have two shifts you get 200 readings per day—if three shifts, 300 readings per day.

3..To conduct the study, pick a spot on the shop floor where you will stand for each machine observation. Walk up to that spot and make an instant observation. If there is some question of what is happening, then follow up with a closer observation or discussion for your recording.

4. You need to inform the shop hands ahead of time of what is coming so you are not being deceptive. If you have a union, of course tie them in from the start.

5. Perform a new delay study every three-to- six months to measure your progress in improving shop efficiency.

Typical readings will include:
- Machine is cutting chips.
- Operator is talking with NC programmer.
- Operator is talking with supervisor.
- Operator is clocked in to the job, but machine is not cutting chips, and operator is absent.
- Machine is broken down—an unplanned breakdown.
- Operator is cleaning up—machine not cutting.
- Operator is performing a setup operation—machine not cutting.

Figure 19-1 will help you decide how many readings to take. For example, 300 readings would be enough for good accuracy. If you only have two shifts, you would probably take 400 readings over two days.

155

Chapter 19

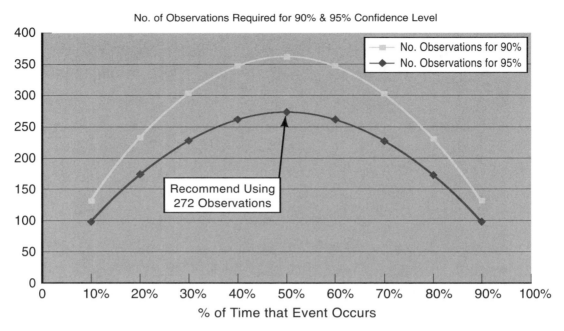

Figure-19-1: Quantity of Observations for Given Confidence Levels

Results of one of my delay studies

- Making chips about 40% of the time.
- Operator absent 19% of the time.
- Setup 15% of the time.
- Tool changes 5% of the time.
- Clean-up 4% of the time.
- Talking 6% of the time

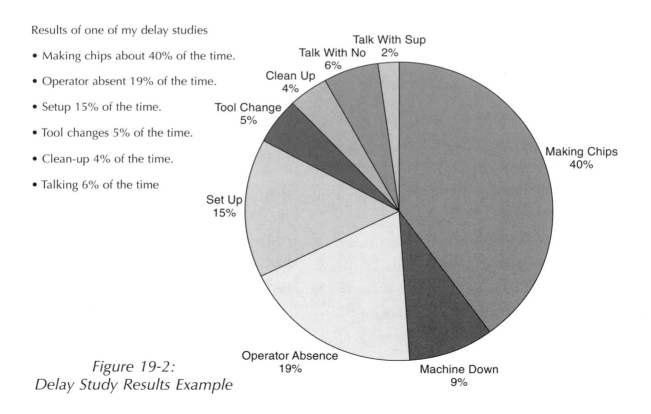

Figure 19-2:
Delay Study Results Example

Measuring Shop Performance

What To Do With the Data

Reduce delays by one half to double machining throughput and halve part-milling costs.

1. Reduce flow time by one half. Use standard tooling bases, load parts off machine, and leave parts on bases through shop layout to verify NC paths. Engineer fixtures to take roughing loads of 6000 pounds (2722 kg) in Z axis. Adjust tool holder pull-out loads to 6000–8000 pounds (2722–3629 kg).
2. Study operator "idle and absent" cases to determine where we can help and implement improvements. Have an industrial engineer or other person follow along with operators for a week or so to see what activities are happening when the operator is absent from the machine. These absences are usually very legitimate.
3. Work with maintenance to reduce machine down time, employ preventative maintenance, use faster communication, implement "response–stocking" of high usage spare parts, and determine when to replace machines.
4. Improve chip containment and covering of "T" slots and frequency of chip cleanup.
5. Help solve the shop load problems.

When you have a picture of the delays many actions for improvement will be obvious.

Speed Up the Metal Cutting Time

Speed up the milling itself—coolants, cutting tool geometry, coating materials, rigidity and speed of machines, and more efficient NC programs. I refer you back to the chapters on milling.

Chapter 19

The Value-Added Operations:

In the old days (the mid 1980s) most of our efficiency efforts and research were aimed at improving the time that it takes to make chips—the value-added time. Today in our lean studies, we tend to focus most heavily on the non-value-added areas (areas we used to call delays).

Sharpening a wooden pencil with a knife is value added while the shavings are being made—should we leave this operation alone? We should only consider leaving value-added operations alone if they employ today's latest technology (or latest affordable technology).

Summary

Many labor collection systems are not working well or give results too late for corrective actions to take place.

Delay studies may be a good and frugal answer for measuring shop performance on a rather frequent basis. These studies can point the way to shop improvements.

Chapter 20

WORK CELLS

This chapter contains some personal observations on industry trends toward "work cells." Then, a few years later, away from "work cells." I discuss the ins and outs of assembly lines, and specialized groupings of machines. With each study, industry makes changes to save money and to improve quality. A bit of the history that led up to the initial placement of machines is covered, on up through today's minor stampede into lean studies. The essence of the chapter is a challenge to weigh proposed improvements carefully and to not make changes just to conform to the "in-crowd."

The manufacturing objective has always been to **continuously stay ahead of the competition and make good profits at the same time,** or words to that effect. Companies are constantly looking for fresh ways to meet this objective; and from my research and personal experiences most fresh approaches work. One conclusion from the Westinghouse studies[3] is that any time we pay special attention to a work group we get improved results.

The focus of this chapter is not on final assembly lines, but on the evolution of detailed part and sub-assembly fabrication that supplies final assembly lines.

---

3 Some controversy has been introduced: "The Westinghouse studies recently criticized (Bramel and Friend, 1981) for yielding erroneous conclusions, had originally and for many years suggested that employee productivity could be improved by increasing the amount of attention accorded to company employees."

Chapter 20

Who knows where it all started, but at some early point machines were human powered. A bit later, animals were harnessed and led in circles to create rotating power. More sophistication was then introduced with waterwheels; they followed animal power). As waterwheels evolved, they became strong enough to supply power to several machines.

Fortunately, an example of an early waterwheel-powered factory has been preserved. I recently visited the birthplace of precision milling—a musket factory in Windsor, Vermont. The Windsor factory was built in 1846. A portion of the original factory that produced the first muskets with interchangeable parts still exists. Heretofore all parts were hand fitted and were not interchangeable. The Windsor factory is now the Precision Museum (see Figure 20-1). This factory and other early factories had a single driven shaft that usually extended the entire length of the factory and provided power for all the machines. The machines were, of necessity, lined up under this one shaft. Belts hung over the common shaft drove each machine. Machines were started and stopped by slipping the belt on or off the pulley of the individual machine. Various diameters of belt pulleys (stepped pulleys) allowed the operator to select a "best" speed for his operation.

Henry Ford changed the world with his assembly lines for final assembly of automobiles and trucks in the early 1900s, but the fabrication of detailed parts and sub-assemblies was not generally accomplished on moving lines.

Transformation to Work Cells

My first direct factory experience was in 1953 at a truck plant in Pontiac, Michigan. In the old machine shop, three long shafts extended the length of the shop and powered all machines. The shafts were overhead like the earlier waterwheel powered factories, but the shafts were now driven by coal-fired steam engines. I was privileged to see history being made, as the individual machines were each re-fitted with their own electric motors, and the last overhead shaft came to a halt.

Then one day a visitor asked "Why are all the machines in straight lines?" This

Work Cells

Figure 20-1: Robbins & Lawrence Armory & Factory, Windsor, Vermont
(photo by the author)

question triggered a study of drill-mill-lathe-grind type of sequences for all our parts. Process and routing sheets for the great variety of parts being machined and of their respective production volumes were studied. Machines were re-arranged into logical "work cells". A typical cell contained combinations of machines such as:

- Mill – drill – mill
- Mill – mill – lathe – drill
- Mill – grind – drill

Chapter 20

Thus "work cells" were born in this shop and solid savings were made. Another savings was the switch to rubber drive belts. Three "belt dresser" jobs for maintaining leather belts were eliminated. Parts completed machining in these new cells.

Demise of Early Work Cells

The cells worked great for about two years, but product part mix and designs changed. Our cells became progressively less efficient and no longer made sense because we were not very flexible. Fresh cost-saving studies were made. The studies suggested that quality and costs could be improved by grouping like machines (and like skills) together. So a lathe department, a grinding area, a milling group, and a drill area were established. When we grouped like machines together, great savings were realized because we had similar skills grouped, like maintenance problems etc. Here all the skilled operators of grinders were able to help each other and easily exchanged ideas on their specialty. Grinding wheels were stored in one place, close at hand. Similar savings were realized for the lathe, mill, and drill areas.

The Beginnings of Assembly Lines

Later, conveyor lines were introduced to handle the movement of parts from area to area. The lines were extended to carry parts through cleaning, then prime and finish paint. A further line extension allowed sub-assemblies to be completed, all on assembly lines. These lines often fed the final assembly lines.

Work Cells Re-emerge

A few years later we read about overseas progress of creating "work cells." We followed the lead of others. With "work cells," we saved inventory by having less material in work. Quality was improved because the workers had the pride and responsibility of completing parts including inspection in individual work cells. Some large parts were cell produced like low-volume trucks, and in other factories, aircraft. Material and parts were fed to the work cell and a complete product was made. There was also productive competition between cells. Costs and quality improved again.

Flexible Work Cells

An SME friend suggested I call this a re-configurable factory rather than flexible work cells.

Flexibility: The ideal shop will have solid concrete floors at least 6 inches thick, and provisions for quickly moving light equipment to form efficient work cells as required. This flexibility can be most efficient in a shop with a thick floor, machines that do not need foundations, and machines that do not need to be bolted to the floor. When I designed a factory a few years back, we hung all electrical lines with receptacles just above head level. Every machine in the shop used 440-volt 3-phase current and every machine and outlet were phased so the machine motors would rotate in the proper direction regardless of where plugged in. All air lines were routed with only one kind of fitting: tees. No couplings or elbows were used. This allowed us to come back in later years to add extra air drops without major expense. Using all tees is not an extra cost because elbows, tees, and couplings all cost about the same and the initial labor for installing them is about the same as for other fittings.

Now with solid floors and useful grids of air and electrical outlets, equipment can be re-arranged into useful work cells in a matter of minutes without the time and expense of bringing in electricians and pipe fitters. One additional feature was to run a capped air line out through the wall of the factory to allow fast hook-up of a rental air compressor in the case of air equipment breakdown. This also keeps the noise of the rental unit outside the building. Another way of keeping costs down is to add accumulator tanks at locations where heavy air draws occur. This allows you to use smaller (less costly) air supply lines.

From a discussion at the annual SME Technical Summit held in 2005 at Oconomowok, Wisconsin, came the following four additional ideas:

Chapter 20

- Modify your equipment so it is easy to move and relocate.
- Use step-down transformers so machines with lower power requirements could be plugged into standard electrical drops. These transformers could be either built in to the machine or could be portable units.
- Locate fixture and tool builders close to the machine shop floor so they are more accessible.
- Design universal containers or parts handlers.

Today We Have Lean Studies

With lean, there appears to be a minor stampede to put every job on a progressive moving or pulsing line.

We have shifted from cells to assembly lines and back again throughout the years: each time we saved time and money and boasted about improved quality. I predict that winning companies will be those who carefully weigh each improvement and do not make changes just so we conform to the "in-crowd."

GROUP VII: ED'S TOOL KIT

Tidbits, and My Tales—Memorable Happenings

Here is a collection of safety items, rules-of-thumb, technical information, happenings, and lessons learned that I use daily as tools of my trade. I enjoy working on shop problems and these are some of my tools. *Chapter 21: Tidbits* is a collection of useful information for working with titanium alloys. Chapter 22: is a collection of observations that were memorable enough for me to record. Many of these notes are of unexpected happenings that help me understand some of the events that I experience in the machine shop world.

My grandfather, Fred Rossman, a master carpenter, helped me build this toolbox in 1942, when I was seven. His words, "Measure twice and saw once," were not invented by him, but have stayed with me over the years. By the way, I asked him why he threw some of the nails away when we were building the box. He answered, "The heads were on the wrong end."

CHAPTER 21

TIDBITS AND RULES OF THUMB

This chapter is intended to be a living document—to be updated on a daily basis as I collect tidbits of information relating to the fabrication of titanium alloy parts.

It has taken me a lot of research and time to find the answers to questions about titanium alloy milling over the past fourteen years. Part of this expensive research effort is because most of us did not grow up with fathers or industries that machined titanium alloys—it was not a daily table topic.

The purpose and scope of this chapter is to list some tidbits and rules of thumb that I have found useful in working with titanium alloy processing. The various tit-bits are listed as one-liners and are grouped under the following headings:

- Safety issues
- Rules of thumb
- Potential resources for information research on Titanium alloy Milling

Safety Issues

Titanium burns and it is tough to put out a titanium fire. Water does not put it out, but water may slow the spread of fire through its cooling action. Experts put the fire out by smothering it with chemicals. I have seen a small fire put out with sand. My suggestion, if the fire is portable, like in a tote box full of chips, is to get the box

Chapter 21

outside the building and let it burn out. I have seen two fires handled in this way by alert forklift drivers.

Preventive actions are to keep chips cleaned up and keep them damp at all times. Remove all possible chips from the shop. Put the boxes outside—there is no prize for neatness by not storing chips outside. Besides they don't rust.

Rules of Thumb

- Six in.^{3} of titanium alloy weighs one pound—almost exactly (220 cm^3 ~ 1 kg)
- When milling, 1.4 horsepower removes 1 in.^{3}/min. (0.016 dm3/min); therefore, high horsepower is not needed. A good roughing operation removes about 5 in.^{3}/min. (0.082 dm^3/min). Remember that cutting torque approximately doubles as the cutter wears.
- Horsepower, however, is a good indicator of machine rigidity.
- The Coefficient of Thermal Expansion CTE of titanium alloy ti-6Al-4V-ELI is 5.7 x 10^{-6} in./in./°F (11.0 x 10^{-6} mm/mm/°C).
- Mild steel is a great tooling material for most alloys of titanium with a CTE of about 6.6 x 10^{-6} in./in./°F (11.9 x 10^{-6} mm/mm/°C) (almost the same expansion as the titanium alloy being milled).
- Melting point of ti-6Al-4V-ELI is about 3100°F (1704°C).
- Most property changes and formation of Alpha-Case (titanium rust) do not happen below 1000°F (538°C).
- During beta-anneal both ti-6al-4V and ti-6Al-2Sn-2Zr-2Cr-2Mo-STA shrink about 0.0007 in./in. (0.0178 mm) of length. This shrinkage appears to be rather consistent because it even happened when I hung a spar by the top end during beta-anneal. Here the shrinkage overcame the pull of gravity.

Tid-Bits and Rules of Thumb

Regarding **shrink-fit tool holders**: we have to be very careful to eliminate slippage between cutter and tool holder.

Buy all carbide cutters with a matt finish on the shank.

Stay away from centerless ground stock because it is not round enough (slight oval.) Insist on stock that is ground between centers.

Hold a close tolerance to keep the gap between holder and tool in the 0.0001–0.0003 in. (0.0025–0.0076 mm) range at insertion temperature.

For long milling reaches with cutters, make up telescope-like groupings of heat-shrink tool holders: these are cheaper than specially ground "tapered tool holders."

Poor-mans temperature control: Put temperature control on the machines coolant, then use heavy flood of coolant all over the part. This fire hose effect wins the battle for controlling part and fixture temperature. Controlling coolant temperature is cheaper than the cost of air-conditioning. This idea came to me when I saw hot coolant negate the efforts of air conditioning to control part temperature during some titanium alloy milling tests.

Potential Resources for Information Research on Titanium Alloy Milling

- Big companies like Boeing, Lockheed
- Major Tier 1 suppliers
- SME technical papers and advisors
- Non-profit organizations such as Manufacturing Extension Partnership (MEP)
- Small machining companies
- Government Research Organizations such as Wright Patterson

Chapter 21

- National Center for Defense Manufacturing and Machining
- Cutter manufacturers
- Machinery manufacturers
- Colleges
- Private research organizations
- Web sites
- Consultants

CHAPTER 22

MY TALES—MEMORABLE HAPPENINGS

Here are four tales of unusual shop happenings. The message is that the unexpected does happen. You will be reminded to keep a cool head when unusual events occur, and you might be able to use some of these thoughts in trouble shooting, especially from the first example.

Hot Equipment Quits Working

Tally Corporation, ?1978, we experienced sporadic machine failures. Machines would quit working then, when tried later, they would often work again for a while then, stop again. We finally tracked down the problem. The stoppages happened on hot days when shop temperatures exceeded 90^0 F (32^0C). Certain electrical components would not function properly at high temperatures. The problem was very elusive for a while. Remember that electronics are sensitive to high temperature and can drive you nuts with their antics.

Minds Have Great Power

We introduced a new titanium alloy ti-6Al-2Sn-2Zr-2Cr-2Mo-STA into the machine shop for milling. The word on the street was that this new alloy was particularly tough to machine and that was the reason it had not been used until now. Boeing, 1993, I received a frantic call from the machine shop about tool breakage on the first attempts to mill the newly-introduced ti-6Al-2Sn-2Zr-2Cr-2Mo-STA. The superintendent had called my boss and wanted Manufacturing Research and Development support immediately.

I went to the shop and worked with feeds, speeds, and chip loads. We soon had the problem under control with normal tool life—normal meaning life equivalent to that achieved when cutting ti-6Al-4V-ELI-BA. We had been cutting ti-6Al-4V-ELI-BA successfully for about 30 years, but this new stuff was "impossible to machine." As I rounded a corner headed back to my desk, I noticed a block of material on a pallet that looked much like the trouble part on the mill. On closer examination, the material on the pallet was the block of new ti-6Al-2Sn-2Zr-2Cr-2Mo-STA. A further check revealed that the "un-machineable" material on the mill was, in fact, the old ti-6Al-4V-mill annealed. The shop had inadvertently loaded the wrong titanium alloy onto the machine. The material that they could not mill was in fact the same material that they had been milling for 40 years. Everyone settled down and we switched to the new exotic material and had no problems with it. Is this mind over matter or what?

Sometimes We Jump Too Soon

Tally quit producing punched-hole tape too soon. When magnetic tape and discs began to show up in manufacturing, we quickly realized that our product was soon to be obsolete. In a matter of a half-year, we phased out our old products that ran machines with one-inch wide paper tape with punched-hole patterns. The new magnetic tapes and discs did not catch on for about another two years, and we were caught without product to sell. One of the reasons for the delay in accepting this new product was the skill of the old operators who could see and read the hole patterns in the paper tape. They were untrusting of the new magnetic tapes because there was nothing to see on it. I liken this to the new detergents that came out about 30 years ago that worked great but did not have any suds or bubbles—housewives would not buy it because they could not see any soap suds.

Sometimes We Outsmart Ourselves

In our quest for high-speed milling of titanium alloys we tested carbide cutters, but did not test cobalt cutters because we knew they would not work. Wrong. Our testing showed successful cutting with cobalt (M42) cutters at 400 sfm and radial engagement of 0.025 in. (0.635 mm) to confirm the supplier results.

INDEX

B
balancing 119, 122
ballbar 46
benching 42, 105, 107-11, 145

C
carbide cutters 15, 18, 62, 67, 69-71, 73-7
CATIA 2, 14-5, 21
chatter 4, 27, 30-1, 34, 63-4, 67, 71, 91, 132
check list 9
clamped-in stresses 98-9, 101-2
climb milling 22, 62, 70, 73, 77
cobalt cutters 15, 62-5, 68-9, 73, 75-6, 90, 121, 133, 146, 172
controllers 14, 28, 42, 56
conventional milling 22, 62
coolant temperature 50-1, 80, 145, 169
CTE (co-efficient of thermal expansion) 37, 47, 49-50, 52, 84, 168
cutter burns 89-90, 131
cutter checking 3, 117, 123-5, 127
cutter coatings 5, 115, 120, 129, 138-9
cutter life 3, 7, 30, 56, 63-7, 70-1, 73, 75, 77, 90, 117, 120, 122, 127, 129, 130-3
cutter wear 68-9, 90, 124, 126-7, 131, 168
cutters 4, 5, 6, 11, 15, 43-5, 62-9, 70-1, 73-7, 89-90, 100, 102

D
delay study 12, 153-6
determinists 45-9
dovetail 27, 31
draw bar 17, 34

drawbar pull 34
drill jigs 51, 81-5

F
fail safe pins 3, 13, 124-7
five-axis 29, 32, 39-40, 42-3, 45, 51-2, 54-5, 58, 74, 81-2, 84, 86
five-axis machines 42, 52, 54, 84, 86

G
glass scales 52

H
heat-shrink tool holders 122, 169
heat-shrink 122, 169
high-speed milling 54, 62, 75, 145, 172

I
inherited stresses 98, 101

L
laser 3, 46, 51-2, 84, 125-7
laser tracker 46, 52
Lean Manufacturing 2, 21, 137-8
lean studies 5, 22-3, 61, 135, 139, 141, 149, 151, 158-9, 164

M
machined-in stresses 98, 100, 102
mild steel 37, 50, 84, 168
multi-spindle 29, 54-5, 126

Index

P
plunge roughing 67-8, 71, 140, 146
pogo tool 31-2
powdered metal cutters 67, 70-1
preventative maintenance 3, 11, 57, 157
process capability 47-9, 53, 81, 86

R
riser tables 55-6
rotary fixture 28-30
rough surface 61, 89, 95-6

S
sanding 42, 105, 107-11
share technology 4, 6, 59
spindle run-out 11, 13, 57, 123
spring passes 93, 95
statisticians 45
symposium 5, 6, 139-41

T
thin webs 61, 89, 91
three-axis 39, 45, 54, 107
through-the-spindle coolant 57, 68
Ti-6Al-2Sn-2Zr-2Cr-2Mo-STA 62, 70, 100, 168, 171-2
Ti-6Al-4V-ELI-BA 37, 62, 70, 100, 131, 146, 168, 171
TiN 146
titanium burns 2, 167
tool holders 11, 16, 119-22, 169
tooling balls 35-6, 49
tooling tabs 27, 33, 93
TPM (total preventive maintenance) 2, 23
training 1-2, 12-3, 17, 19, 20-1, 23, 41, 138, 140, 154
true position 43, 52, 79, 80-1, 86-7

U
Unigraphics 2, 14, 21

V
Vericut 14-5, 21-2
vibratory finishing 105, 111, 113-6
volumetric accuracy 49

W
warpage 20, 28, 61, 92, 97-9, 100-3, 145
wave cutters 15, 66
work cells 2, 3, 11, 135, 159-63

X
x.ceed 77